Lee

Blow Molding Design Guide

SPE Books from Hanser Publishers

Belofsky, Plastics: Product Design and Process Engineering
Bernhardt, Computer Aided Engineering for Injection Molding
Brostow/Corneliussen, Failure of Plastics
Chan, Polymer Surface Modification and Characterization
Charrier, Polymeric Materials and Processing—Plastics, Elastomers and Composites
Del Vecchio, Understanding Design of Experiments: A Primer for Technologists
Ehrig, Plastics Recycling
Ezrin, Plastics Failure Guide
Gordon, Total Quality Process Control for Injection Molding
Gordon/Shaw, Computer Programs for Rheologists
Gruenwald, Plastics: How Structure Determines Properties
Lee, Blow Molding Design Guide
Macosko, Fundamentals of Reaction Injection Molding
Malloy, Plastic Part Design for Injection Molding
Manzione, Applications of Computer Aided Engineering in Injection Molding
Matsuoka, Relaxation Phenomena in Polymers
Menges/Mohren, How to Make Injection Molds
Michaeli, Extrusion Dies for Plastics and Rubber
Michaeli/Greif/Kaufmann/Vossebürger, Training in Plastics Technology
Michaeli/Greif/Kretzschmar/Kaufmann/Bertuleit, Training in Injection Molding
Neuman, Experimental Strategies for Polymer Scientists and Plastics Engineers
O'Brien, Applications of Computer Modeling for Extrusion and Other Continuous Polymer Processes
Progelhof/Throne, Polymer Engineering Principles
Rauwendaal, Polymer Extrusion
Rees, Mold Engineering
Rosato, Designing with Reinforced Composites
Saechtling, International Plastics Handbook for the Technologist, Engineer and User
Stevenson, Innovation in Polymer Processing
Stoeckhert, Mold-Making Handbook for the Plastics Engineer
Throne, Thermoforming
Tucker, Fundamentals of Computer Modeling for Polymer Processing
Ulrich, Introduction to Industrial Polymers
Wright, Injection/Transfer Molding of Thermosetting Plastics
Wright, Molded Thermosets: A Handbook for Plastics Engineers, Molders and Designers

Norman C. Lee

Blow Molding Design Guide

Hanser Publishers, Munich
Hanser/Gardner Publications, Inc., Cincinnati

The Author:
Norman C. Lee, 2705 New Garden Road East, Greensboro, NC 27455, USA

Distributed in the USA and in Canada by
Hanser/Gardner Publications, Inc.
6915 Valley Avenue, Cincinnati, Ohio 45244-3029, USA
Fax: (513) 527-8950
Phone: (513) 527-8977 or 1-800-950-8977
Internet: http://www.hansergardner.com

Distributed in all other countries by
Carl Hanser Verlag
Postfach 86 04 20, 81631 München, Germany
Fax: +49 (89) 98 12 64

The use of general descriptive names, trademarks, etc., in this publication, even if the former are not especially identified, is not to be taken as a sign that such names, as understood by the Trade Marks and Merchandise Marks Act, may accordingly be used freely by anyone.

While the advice and information in this book are believed to be true and accurate at the date of going to press, neither the authors nor the editors nor the publisher can accept any legal responsibility for any errors or omissions that may be made. The publisher makes no warranty, express or implied, with respect to the material contained herein.

Library of Congress Cataloging-in-Publication Data
Lee, Norman C., 1934–
Blow molding design guide / Norman Lee.
   p. cm.
Includes bibliographical references and index.
ISBN 1-56990-227-5
1. Plastics—Molding. I. Title.
TP1130.L44   1998
668.4'12—dc21   97-43193

Die Deutsche Bibliothek – CIP-Einheitsaufnahme
Lee, Norman:
Blow molding design guide / Norman Lee. – Munich : Hanser ; Cincinnati : Hanser/Gardner, 1998
   (SPE books)
ISBN 3-446-18255-1

All rights reserved. No part of this book may be reproduced or transmitted in any form or by any means, electronic or mechanical, including photocopying or by any information storage and retrieval system, without permission in writing from the publisher.

© Carl Hanser Verlag, Munich 1998
Typeset in England by Techset Composition Ltd., Salisbury
Printed and bound in Germany by Kösel, Kempten

Foreword

The blow molding industry continues to grow and become more sophisticated; however, there is little written about basic blow molding design. The *Blow Molding Design Guide* offers a spectrum of design information useful for someone new in blow molding, learning the basics, as well as someone experienced in blow mold design looking for new approaches to resolve difficult part design issues. The list of references throughout the book is a virtual who's who in the blow molding industry.

Norman Lee has more than 25 years of experience as a design engineer in the blow molding industry. He is well known for his years of service to the plastics industry through leadership in the Society of Plastics Engineers Blow Molding Division and the Plastics Recycling Division. Mr. Lee has demonstrated his commitment to education by service on the SPE Education Committee and by instructing numerous courses in blow molding. Norman Lee was the editor for the first edition of the *Plastic Blow Molding Handbook*, still in print and a resource for many of his classes.

We believe that the *Blow Molding Design Guide* will fill a void in the blow molding industry by condensing a large collection of industry design knowledge into a comprehensive, usable text and reference.

Lewis Ferguson
Chairman
Blow Molding Division
Society of Plastics Engineers

Technical Volumes Committee
Robert C. Portnoy, Chairperson
Exxon Chemical Company

James P. Parr, Reviewer
Exxon Chemical Company

Preface

Designing plastic blow molded parts can be an extremely difficult task because of the complexities of part geometry and the molding processes. It is challenging to even experienced designers. To produce an effective design it should be:

Functional and achieve the objective for which it is intended,
aesthetically pleasing, within the limits of the process,
practical, utilizing the right material, maximizing the benefits of the process,
cost efficient, with consideration of tooling cost and methods and run quantities.

Therefore, the objectives of this book are to give the reader an understanding of plastic blow molding, materials and processes, thus enabling him to design a blow molded part that optimizes the effectiveness of the plastic materials used, process employed, as well as the function of the part. It includes the application of bottles, industrial and structural parts. It is intended to be a no-nonsense, practical hands on book, that forgoes a scientific language that most ordinary people do not understand and concentrates on real life, day to day problems faced by those working to create cost-effective blow molded parts. It is a good introduction to the overall picture, for those who then wish to delve into more detailed and academic aspects of anyone of the many processes discussed.

Because the work includes so many diverse subjects it is not possible for one individual to be an expert in them all, thus, I have relied heavily on experts in their relative fields for information and advice. These of course are acknowledged at the end of each chapter. In many cases I have drawn from the published work. I make no apologies for this, since I am not able to improve on the original work. Also, much of the matter presented is leading edge technology and development by the originators.

I also acknowledge several who have helped me put this manuscript together. Loretta Lee, graduate student at A&T State University, a UNC System in Greensboro, NC, who took it on as a Masters project. Dr. Brent Strong, Brigham Young University, reviewed the manuscript, corrected errors, and made suggestions for changes. The review committee of the S.P.E blow molding division headed by James Parr, Exxon Company, Robert Gilbert, Equistar; and Robert Read, Dow Brands, made valuable suggestions for improvement. Several diagrams were drawn by Sam Huffine, Huffine Associates, Greensboro, NC and Auto-CAD drawings by James Lee, Ashboro, NC.

Norman Lee

Contents

Chapter 1
The Basics of Blow Molding
1

1.1 Definition	1
1.2 The Basic Process	1
1.3 History	2
1.4 Summary of Development	2
1.5 Design Parameters; Benefits, Disadvantages, and Comparisons	8
1.6 References	11

Chapter 2
The Design Process – An Organized Approach
13

2.1 Introduction	13
2.2 Main Structure	13
2.2.1 Six Perspectives on the New Manufacturing Enterprise	14
2.3 Product Design and Development Management System (PD2MS)	16
2.3.1 PD2MS	16
2.3.2 Process Management Tracking System	17
2.3.3 Commitment of Resources	17
2.3.4 Concurrent Engineering	19
2.4 Conclusion	19
2.5 References	25

Chapter 3
Basic Blow Molding Part Design
27

3.1 Basic Design Considerations	27
3.1.1 Size Variations	27
3.1.2 Understanding Hollow Structures	27
3.1.3 Draft of Part	29
3.2 Increase Draft as Blow Ratio Increases	30
3.3 Guidelines for Radii	33
3.3.1 Corner and Edge Rounding	35
3.3.2 Chamfers	36
3.4 Molded-in Geometric Configuration	36
3.5 Flanges and Tack-offs	37
3.5.1 Threaded Parts	38
3.5.2 Attachments and Auxiliary Units	39
3.6 Conclusion	40
3.7 Reference	41

Chapter 4
Design for Bottles
43

4.1 Blow Molding Process Basic Shapes.	43
4.2 Simplified Assumptions About Parison Expansion.	44
4.3 Bottle Design Concepts.	44
4.3.1 Ribs: Do Not Always Stiffen	46
4.3.2 Cross-Sections	47
4.3.3 Bottle Neck, Threads, and Openings	48
4.4 Container Volume.	48
4.4.1 Container Volume Measurements	51
4.4.2 Standards	52
4.4.3 Machine Line Mold Volume Correction	54
4.4.4 Source of Error in Volume Correction	54
4.4.5 Package Dairy Mold Volume Correction	54
4.4.6 Production Conditions.	55
4.4.7 Conclusion	56

Chapter 5
Industrial and Structural Part Design
59

5.1 The Blow Molding Process	59
5.1.1 Preferred Process.	59
5.1.2 Hollow Parts	59
5.1.3 Resin/Fiberglass Lay up and Structural Foam Molding	60
5.1.4 Foam-Filled	60
5.2 Kinetic Energy Design Engineering	62
5.2.1 Energy Management Concepts.	62
5.2.1.1 Deformation (of the Surface Skin in the Area of Impact)	62
5.2.1.2 A Sealed Hollow Part	62
5.2.1.3 Foam-Filled	62
5.2.1.4 Crushing	63
5.3 Molded-In Insert of Components	63
5.4 Interlocking Systems	65
5.5 Snap Fits	65
5.6 Multiple/Combination Cavities	66
5.7 Container Configuration Design.	66
5.7.1 Flat Sides	67
5.7.2 Lip	67
5.7.3 Nesting and Stacking	68
5.7.4 Cutting Containers Apart	68
5.8 Conclusion	70
5.9 Reference	72

Chapter 6
Computer Aided Design and Engineering Analysis
73

6.1 Performance Criteria	73
6.2 Computer Software Simulation	73
6.3 Reducing Parison Thickness.	74
6.4 Fluid Flow Finite Element Simulation.	76
6.4.1 Modeling	76

  6.4.2 Simulation 77
  6.4.3 Prediction Example 78
6.5 Polymer Inflation and Thinning Analysis 80
  6.5.1 Geometric. 80
  6.5.2 Understanding Wall Thickness 80
    6.5.2.1 Using Normalized Thickness Curves. 82
6.6 Conclusion 83
6.7 References 83

Chapter 7
Decorating of Blow Molding Products
85

7.1 Introduction 85
7.2 Surface Treatment. 85
  7.2.1 Surface Treatment Methods 86
  7.2.2 Flame Treatment 86
  7.2.3 Corona Discharge 86
  7.2.4 Washing with Water-Based Chemicals. 88
  7.2.5 Solvent Cleaning and Etching 88
  7.2.6 Additives Compounded into Resins 89
7.3 Spray Painting 89
  7.3.1 Air Atomization and Airless Sprays 89
  7.3.2 Masking. 89
  7.3.3 Vapor Degreasing 90
  7.3.4 Mechanical Abrasion-Sanding. 90
  7.3.5 Chemical Etching 90
7.4 Labels. 91
  7.4.1 Label Application 91
7.5 Screen Printing 92
  7.5.1 Screen Printers 94
7.6 Pad Printing 95
  7.6.1 Pad Equipment. 96
7.7 Hot Stamping 97
  7.7.1 Hot Stamping Foils 99
7.8 Decals. 99
  7.8.1 Advantages of Heat Transfer 99
7.9 In Mold Labeling 100
  7.9.1 In Molding Labeling Equipment 100
  7.9.2 In Molding Labeling Process 101
  7.9.3 "In Mold" Label Molds 101
  7.9.4 Cycle Times 102
  7.9.5 Aesthetics 102
7.10 Conclusion 102
7.11 References 102

Chapter 8
The Blow Molding Process
103

8.1 Extrusion Blow Molding. 103
  8.1.1 Understanding the Extruder 104
  8.1.2 Blow Molding Technique 106
  8.1.3 Continuous Extrusion. 108

	8.1.3.1 Shuttle System	108
	8.1.3.2 Rising Mold	109
	8.1.3.3 Rotary Wheel	109
8.1.4	Intermittent Extrusion	110
	8.1.4.1 Reciprocating Screw	110
	8.1.4.2 Ram	111
	8.1.4.3 Accumulator	111
8.1.5	Coextrusion	113
8.1.6	Introduction to Head Tooling	114
	8.1.6.1 Converging	114
	8.1.6.2 Diverging	115
	8.1.6.3 Tooling Choices	116
8.1.7	Part Weight and Wall Thickness Adjustment	116
	8.1.7.1 Parison Programming	117
	8.1.7.2 Die Ovalization	117
8.2 Blow Pins/Needles		119
8.2.1	Needles	119
8.2.2	Pins/Needles	120
8.3 Injection Blow Molding		120
8.3.1	Injection Blow Molding Process	120
8.3.2	The Injection Blow Molding Machine	122
8.4 Stretch Blow Molding		124
8.5 References		126

Chapter 9
New Applications of Blow Molding Technology
127

9.1 Co-Extrusion Blow Molding of Large Parts		127
9.1.1	Reasons for Coextrusion	127
9.1.2	Typical Structures	129
9.1.3	Intermittent and Continuous Extrusion Blow Molding	130
9.1.4	Methods of Continuous Coextrusion Blow Molding	132
9.2 Three-Dimensional Blow Molding		135
9.2.1	Mold Inclining System and Computer Controlled Mold Oscillating Device	136
	9.2.1.1 The X–Y Process	136
	9.2.1.2 Formed Parts	136
	9.2.1.3 Features of the X–Y Machine	137
9.2.2	Three-Dimensional Technology of Suction Blow Molding	139
9.2.3	Three-Dimensionally Curved Blow Moldings	141
9.3 Hard-Soft-Hard and Soft-Hard-Soft Technology		142
9.3.1	Axial Extrusion	142
9.3.2	Preferred Material Combinations	142
9.4 Long-Glass-Fiber-Reinforced Blow Molding		143
9.4.1	Breakthrough	143
9.4.2	15% Long-Glass Fiber	143
9.5 Blow Molding Foam Technology		144
9.5.1	Advantages	145
9.5.2	Blow Foam Technology Products	146
9.6 Conclusion		147
9.7 References		147

Chapter 10
Understanding the Mold
149

10.1 Main Characteristic of Mold Halves	149
10.2 Mold Materials	151
10.2.1 Cast Aluminum and Beryllium	151
10.2.2 Aluminum Plate	151
10.2.3 Steel	151
10.3 Importance of Fast Mold Cooling	151
10.3.1 Heat Transfer Rate	152
10.3.2 Cooling	152
10.3.3 Cooling Lines	152
10.3.4 Pinch-Off Areas	154
10.3.5 Blowing Pin	154
10.3.6 Internal Cooling	154
10.4 Cutting and Welding Parison (Pinch-Off)	154
10.4.1 Pinch-Off Section	154
10.4.2 Uniform Weld Lines	156
10.5 High-Quality, Undamaged Mold Cavity Finish	157
10.6 Effects of Air and Moisture Trapped in the Mold-Venting	157
10.7 Injection of the Blowing Air	158
10.8 Ejection of the Piece from the Mold	159
10.9 Bottle Molds	159
10.9.1 Neck Ring and Blow Pin Design	159
10.9.2 Dome Systems	161
10.9.3 Prefinished System	162
10.9.4 Unusual Problems	162
10.10 Injection Blow Molds	164
10.10.1 Parison (Preform) Mold	164
10.10.2 Neck Ring Insert	165
10.10.3 The Core Rod Assembly	166
10.10.4 Materials for Parison Cavity and Core Rods	166
10.10.5 Design Details of the Blow Mold Cavity	166
10.10.6 Vents	167
10.10.7 Neck Ring Insert	168
10.10.8 Bottom Plug Insert	168
10.10.9 Die Sets	169
10.10.10 Injection Blow Molding Tooling Summary	170
10.11 Conclusion	170
10.12 References	172

Chapter 11
Computer Aided Design and Engineering for Mold Making
173

11.1 Advantages	173
11.2 Systems and Methods	174
11.2.1 Analytical Personal Computer	174
11.2.2 Minicomputer	175
11.2.3 Network Station Approach	175
11.3 Utilizing CAD/CAM in a Mold Making Organization	176
11.3.1 Engineering Activities	176
11.3.2 Manufacturing Activities	179
11.4 Reference	180

Chapter 12
Polymers and Plastic Materials
181

12.1 Basic Polymer Chemistry.	181
12.1.1 Structure of Matter	181
12.2 Polymers.	182
12.2.1 Homopolymers, Copolymers, and Terpolymers	183
12.2.2 Thermoplastic and Thermoset Polymers.	183
12.2.3 Amorphous and Crystalline	183
12.2.4 Fundamental Properties	184
12.2.4.1 Average Molecular Weight.	185
12.2.4.2 Chain Length Linking.	185
12.2.4.3 Morphology.	186
12.2.4.4 Additives, Fillers and Reinforcing Agents.	186
12.3 Physical Properties.	187
12.3.1 Specific Gravity.	187
12.3.2 Melt Flow Rate (Melt Index)	188
12.3.3 Moisture	188
12.3.4 Hardness	189
12.3.5 Tensile Strength and Properties.	189
12.3.6 Creep	190
12.3.7 Basic Polymer Parameters and Their Effect on Product Properties.	190
12.4 Characteristics for Blow Molding.	191
12.4.1 High-Density Polyethylene	191
12.4.2 Acrylonitrile-Butadiene-Styrene.	191
12.4.3 Polycarbonate	193
12.4.4 Polypropylene	194
12.4.5 Polyphenylene Oxide.	195
12.4.6 Polyethylene Terephthalate	196
12.5 Coloring Plastic Materials	196
12.6 Regrind	197
12.6.1 Regrind Specifications	197
12.6.2 Process Performance	197
12.6.3 Physical Properties	197
12.7 Post-Consumer and Industrial Recycled Materials	198
12.8 References.	199

Chapter 13
Cost Estimating
201

13.1 Introduction.	201
13.2 Typical Cost Sheet.	201
13.3 Cost Conclusion	204
13.4 Cost Estimating Calculations.	204
13.4.1 Part Requirements.	204
13.4.2 Cooling of the Part	204
13.4.3 Throughput of the Machine	205
13.4.4 Post-Molding Operations	205
13.4.5 Setup and Purging of Material from Previous Product Run	205

Index
205

1
The Basics of Blow Molding

1.1 Definition

Plastic blow molding is a process used to produce hollow component parts. It is confined to thermoplastic type resins, for example, polyethylene, polyvinychloride, polyethylene terephthalate, and engineering plastics such as polycarbonate. The three major variants of blow molding to produce blown plastic components are extrusion blow molding, injection blow molding, and stretch molding [2].

1.2 The Basic Process

The production plant for a blow molding process consists of three stages:

1. *Melting and plasticizing.* An extrusion and/or injection machine may be used to produce the melt.
2. *Parison formation* through a head and die/or injection mold.
3. *Blowing and molding.* Auxiliary air compressors provide air and a hydraulic clamp unit holds the mold.

In these major processes the first step involves the production of the tube, widely known as the parison, a term borrowed from the glass industry. The parison may be produced either by an extrusion or an injection machine. In the latter case it is usually referred to as a preform [3].

The heated parison or preform is placed inside the blowing mold, which closes and clamps around it and then the heated tube is blown against the wall of the mold molten plastic, or resin is then set to shape by being cooled, and after this cooling stage the product is ejected. In many cases the product requires subsequent finishing operations, for example, the removal of flashing, printing and labeling, filling the product, etc. Robotic handling equipment may be used to finish parts using boring and milling operations. For the basic process see Fig. 1.1 [4].

2 The Basics of Blow Molding

Figure 1.1 Glass blowing

1.3 History

Glass, plastics, and aluminum are three classes of raw materials existing today which are blown to form molded parts. The techniques of modern blow molding plastics grew from the art of glass blowing (Fig 1.1). The method is attributed to Syrian glass workers in the first century BC, who realized that a glass bulb on the end of a blow pipe could be shaped to many useful hollow forms, with handles and feet and decorated adjuncts added at will. During the Middle Ages, chiefly in Great Britain and elsewhere in Europe, the process was refined and sophisticated, becoming an important commercial industry [1].

1.4 Summary of Development

The modern plastic blow molding process (Fig. 1.2) originated in the 1930s, when the initial patents were granted to Plax Corporation and Owens of Illinois for automated equipment based on glass blowing techniques (Fig. 1.3). But the high cost and poor performance of plastic materials at that time discouraged rapid development. The plastic bottles offered no advantage over glass bottles; however, the introduction of low-density polyethylene in the mid-1940s (developed by ICI of England) provided the advantage of squeezability which glass could not match. In 1950 Elmer Mills was granted a patent for a continuous extrusion rotary blow molder used privately by Continental Can. In the late 1950s high-density polyethylene and commercially available molding equipment were developed and the industry rapidly expanded [4].

1.4 Summary of Development

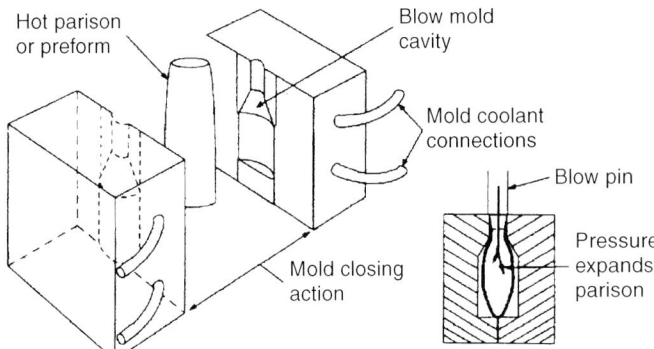

Figure 1.2 Basic blow molding process

Figure 1.3 An extrusion blowmolder developed by Plax Corporation in the mid 1930s for the production of cellose acetate bottles

4 The Basics of Blow Molding

High-density polyethylene added considerable breadth to the design capabilities of plastic bottles; most importantly bottles could be lighter and stiffer. Commercial equipment broadened the opportunities for blow molding. Until that time, blow molding was being performed by only a few companies using proprietary technology. Because of the closely held patent situation in the United States most of the equipment development was taking place in Europe, principally in Germany. The first commercial equipment for blow molding in the United States came from Europe, and was shown at the National Plastic Exhibition in 1958. Empire Plastic, a toy company, bought a Fischer Blow Molding machine to make toy bowling pins. The engineers at Empire also converted a Reed–Prentice injection molding machine to make toy baseball bats by moving the injection cylinder up over the platen and building a head with a die and bushing. (Fig. 1.4). A later step was to build a twin head with a Rotec valve to operate and divert the material flow from one side to another (Fig 1.5). Midland Ross–Hartig gained permission from Empire to use this design to build six machines for making doll bodies for Ideal Toy Company. These were the first blow molding machines built by Hartig, which manufactured extrusion machines. The company later became Waldron–Hartig, then Battenfield–Hartig, and are now owned by Davis Standard. In that same period Paul Marcus designed and built twin head machines called Auto-Blow. During 1960 ZARN Inc. was formed in North Carolina and made milk bottles for Borden Dairy, in conjunction with Uniloy (now Johnson Controls), who built the machines and molds. To eliminate the freight cost of transporting empty bottles, in-line bottle blowing was started by a dairy in Burlington, North Carolina using the Uniloy machines and molds [3].

Of all the plastic materials that can be blown like glass, polyethylene is used in greater volume than all others combined. Although mainly still used for bottles, blow molding is increasingly used for industrial parts such as automotive rear deck air spoilers, seat

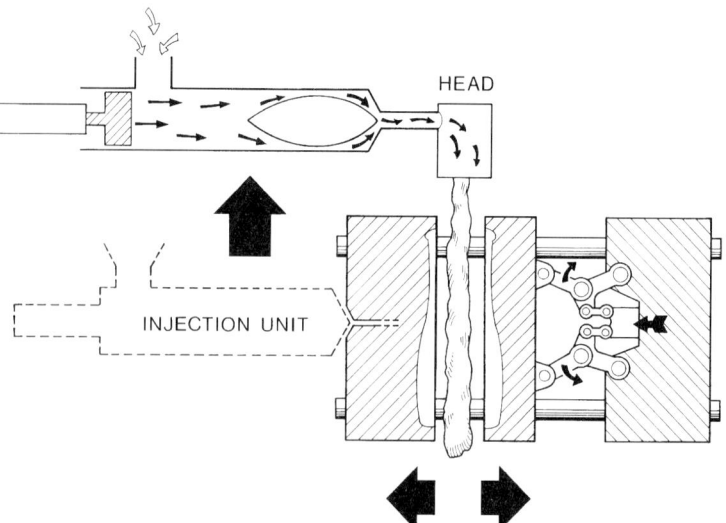

Figure 1.4 Cross-section through converted injection molding machine

1.4 Summary of Development 5

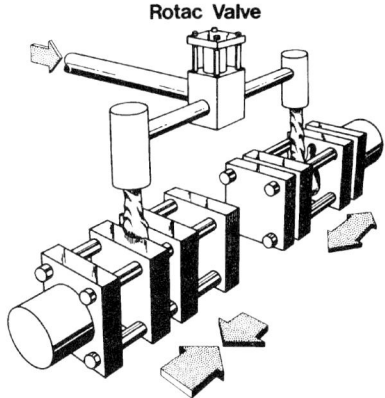

Figure 1.5 Rotac valve

backs, toy tricycles, wheels, typewriter cases, surfboards, flexible bellows, and fuel tanks (Fig. 1.6).

The development of blow molding was fueled by the introduction of high molecular weight polyethylene. Because of its wide ranges in density, melt indexes, and other basic characteristics, this material results in a corresponding wide variation in possible end properties. Flexible bottles are best made from low or medium density polyethylene. High-density plastics are more suitable for rigid parts. The properties common to all items blown from polyethylene include light weight, toughness (even at low temperatures), resistance to attack and penetration from chemicals, resistance to cracking under stress when holding liquids (environmental stress crack resistance), and excellent moldability. Engineered

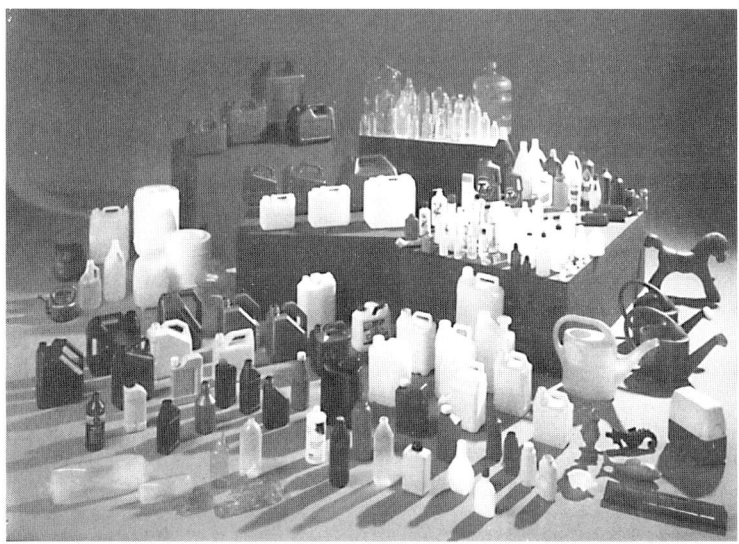

Figure 1.6 Array of blow molding products (Courtesy of Krupp Kautex)

6 The Basics of Blow Molding

plastics, such as PET, PC, and PA, are increasing in use and now make up an important segment of the blow molding market.

Figure 1.7 is a flow chart of the components in the blow molding process. There are probably more differences in equipment for blow molding than for any other plastic fabrication technique. A blow molding machine may be the size of an office desk, making hollow objects as small as a pencil, or it may occupy a large room, making objects as large as 1937.5 l (500-gallon) capacity.

Table 1.1 shows the growth of the blow molding market, in large part attributed to the uniqueness of the process in producing a complete hollow part. This capability cannot be duplicated by injection molding, thermoforming, or metal stamping. Moreover, in

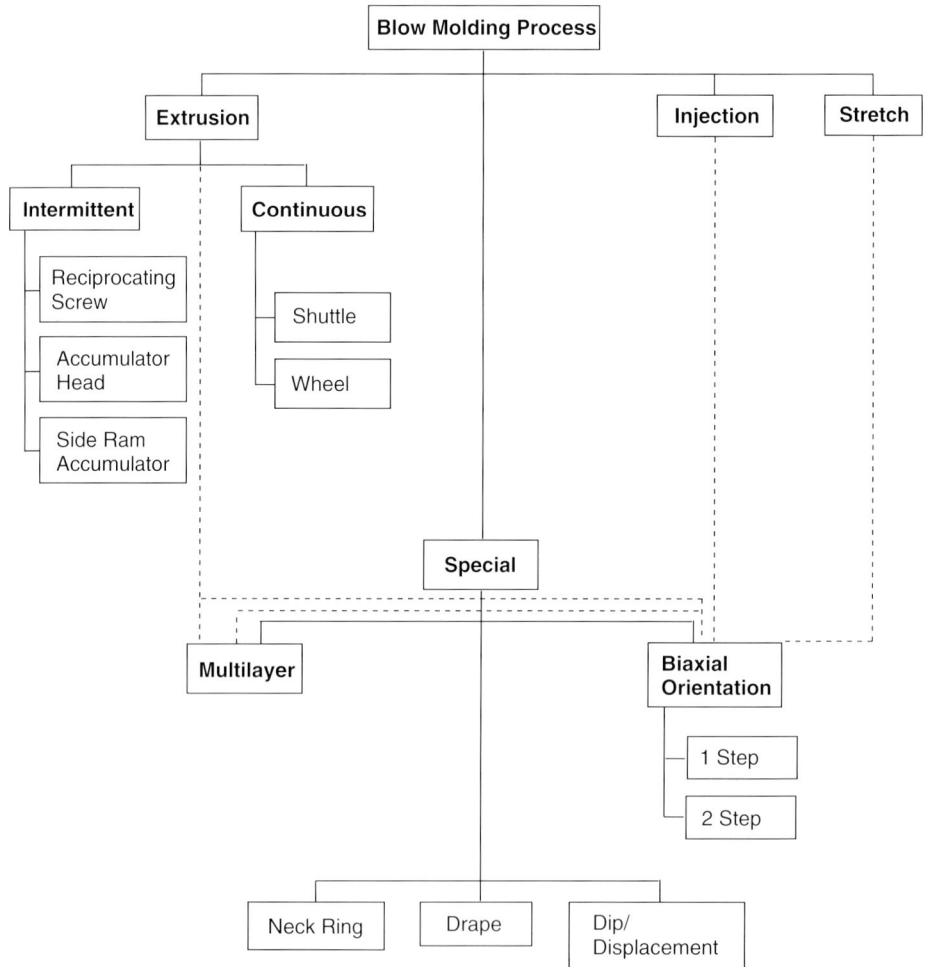

Figure 1.7 Blow molding process

Table 1.1 Growth in the Consumption of Resins, Alloys, and Blends In North American Industrial Blow Molding 1995–2000

	Million lbs 1995	Million lbs 2000	Average annual growth rate (%)
Resins			
Polyethylene	1012	1630	10
HDPE	989	1593	10
LDPE	20	32	10
LLDPE	1	2	15
EVA	2	3	5
Polypropylene	105	185	12
PVC	1	1	0
Polystyrene (HIPS)	5	5	0
Subtotal	1123	1821	10
Engineering resin, alloys, and blends			
Nylon	17	21	4
Thermoplastic elastomers	16	26	10
Thermoplastic polyurethane	5	7	7
Polysulfone	1	2	15
Polycarbonate	0.6	0.6	0
ABS	1	1	0
PC/ABS	1	3	25
PC/PET	1	1	0
Styrene–Butadiene	6	10	11
PPO/PS	15	19	5
PPO/nylon	1	2	15
Nylon/PET	10	14	7
Nylon/ABS	0.4	0.5	5
Others (incl. EVOH)	0	5	–
Subtotal	75	112	9
Grand total	1198	1933	10

comparison to other molding processes, blow molding is less expensive in relation to injection mold tooling cost and assembly cost. Figure 1.8 shows the distributions of resins, alloys, and blends in North America [5].

8   The Basics of Blow Molding

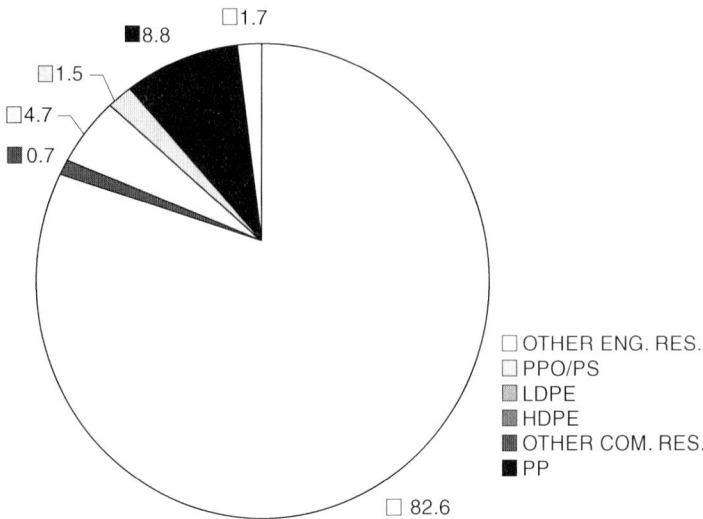

Figure 1.8 Distribution of resins, alloys, and blends in North American Industrial Blow Molding, 1995 (Courtesy of PCRS)

1.5 Design Parameters; Benefits, Disadvantages, and Comparisons

The blow molding process is a natural choice for containers and hollow parts, the only competitive process being rotational molding. In rotational molding a measured quantity of powdered material is loaded into aluminum shell molds that are closed and placed into an oven for heating. The material, tumbles against the hot surface and sticks to them, building thickness and uniformity as the mold turns. The molds are taken from the oven and cooled in a cooling chamber. Because the material is fused from powder particles and does not have a homogeneous structure, the physical properties are generally not as good as those of a blow molded product. It is good, however, for low volume production since low investments are required for mold which are much simpler to make (Fig. 1.9).

Blow molding is the preferred solution in high volume, high strength applications such as automotive and agricultural tanks, pressure vessels, air ducts, as well as wire and cable channels. When high structural strength is required the part may be filled with foam. This method can also be used for applications when insulation or buoyancy is required.

The designer, then, when looking at a part that is being considered for the blow molding process needs to come to terms with the hollow walled structure. In designing for most manufacturing processes, a single-walled shell is usually considered, formed or molded into the desired configuration. The wall thickness is governed by the process practicability, appearance, and economic considerations.

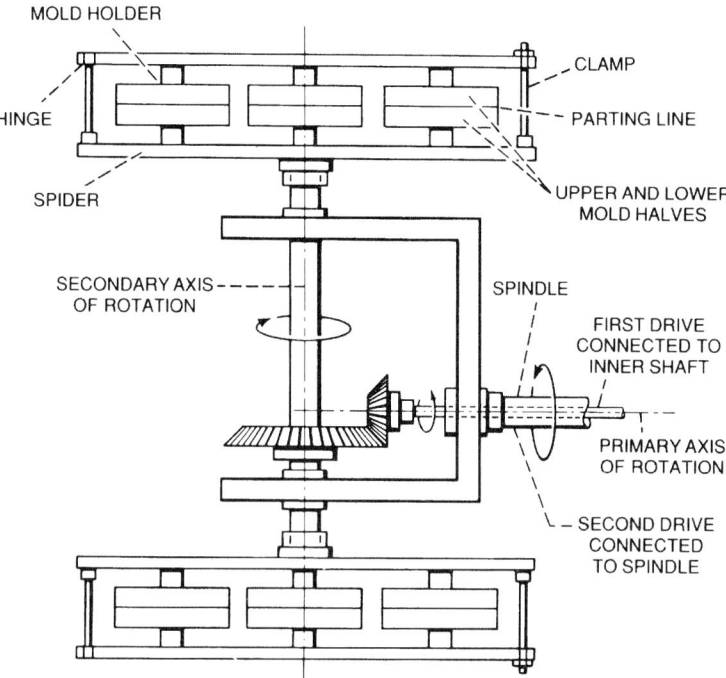

Figure 1.9 Schematic presentation of a typical system used to obtain mold rotation in two planes perpendicular to each other. The spindle is turned on the primary axis (center). Usually, the ratio of the two simultaneous rotations is determined by the gear (center), which may be exchangeable, or, occasionally, by two motor drives being available. A ratio of 4 : 1 usually gives good results. (Courtesy of Millennium Petrochemicals)

Thermoforming, whether pressure or vacuum, can also be used to form hollow parts by forming a plastic sheet into a female cavity or male core, and then welding the two trimmed parts together. Another example would be utilizing the methods used in sheet metal forming and assembling where the strength and stiffness are created by bending, forming, and attaching together by welding or a "pinched" formed seam. Figure 1.10 provides a description of injection molding, rotation molding and thermoforming and compares the advantages and disadvantages of each process.

10 The Basics of Blow Molding

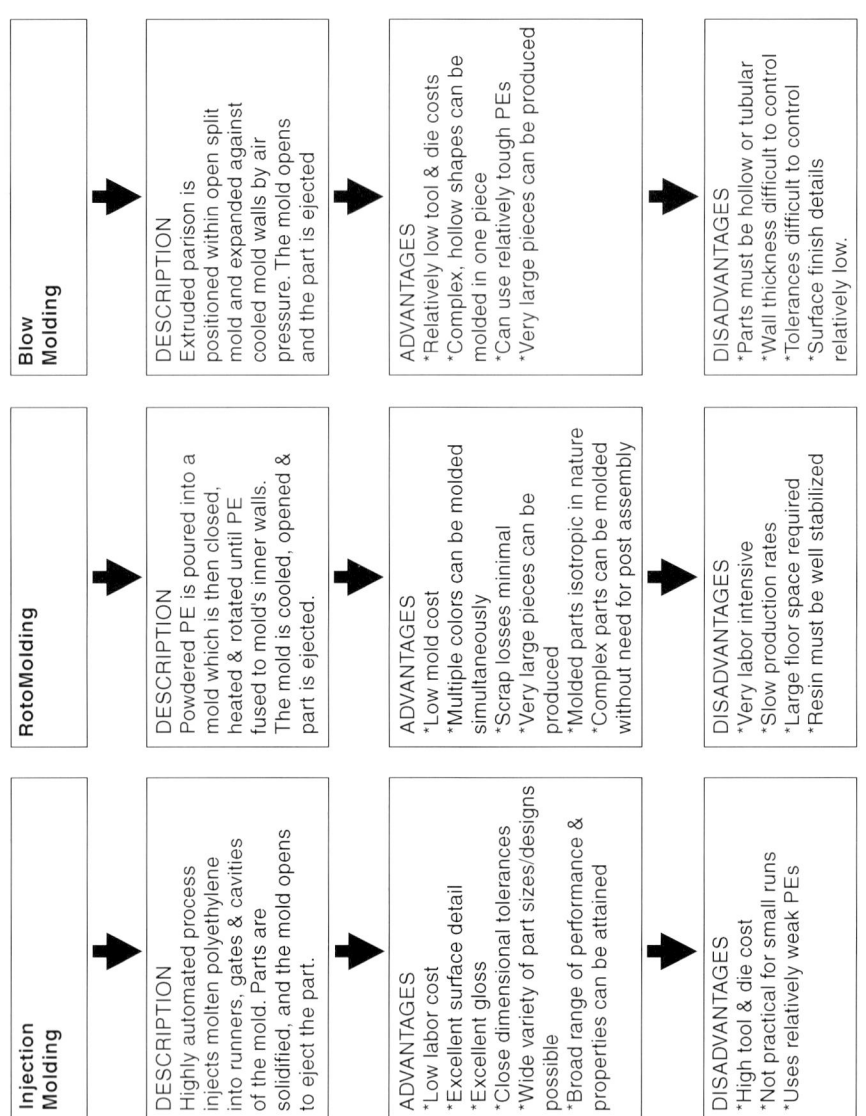

Figure 1.10 Comparison of plastic molding processes

1.6 References

1. Glass. *In Blowing*—Source unknown
2. Quantum USI Division. *Polyolefin Blow Molding and Operating Manual* (now Equistar LP.)
3. Whelan, T. *The Bekum Blow Molding Handbook*
4. Lee, N. (Ed.) *Plastic Blow Molding Handbook*. Van Nostrand Reinhold, New York
5. Mooney, P. Plastic Custom Research Service

General References

1. *Encyclopedia of Polymer Science Engineering*, 2nd edit., Vol. 2. John Wiley & Sons, New York
2. Quantum U.S.I. *Petrothene Polyolefins—A Processing Guide* (now Equistar LP.)
3. Plastic Custom Research Services, Advance, NC. *An Analysis of North American Industrial Blow Molding Business—A New Surge of Growth in the 1990s*

2
The Design Process
An Organized Approach

2.1 Introduction

When designing for plastic blow molded parts, which is a high-volume production process, esthetics, functional process, capabilities, and manufacturing constraints must be considered. The designer should have an overall grasp of these issues (the purpose of this book) but cannot be an expert in all aspects because many disciplines are involved. The key to success is an understanding of the setting—large or small company, division, or client relationship, etc. [1].

Decisions made at the Product Design level have a profound effect on each product cost factor: assembly, maintenance, and total life cycle cost. The relationship among design, materials, and process functional activities is shown in the Venn diagram in Fig. 2.1 [2].

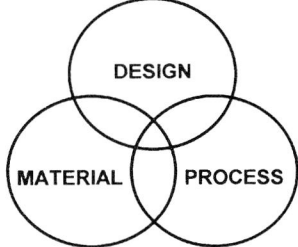

Figure 2.1 Venn diagram (Courtesy of Brigham Young University)

2.2 Main Structure

An employee or consultant working with a manufacturing organization must have the ability to work with people of different disciplines and backgrounds and is also required to select the right materials and processes to achieve the design objective. A method to provide a system to facilitate the team approach was developed at Brigham Young University called the Computer Integrated Manufacturing (CIM) system, modeled in the form of a wheel (see Fig. 2.2) [3]. This was further refined by the Computer and Automated Systems Association of Society of Manufacturing Engineers.

14 The Design Process—An Organized Approach

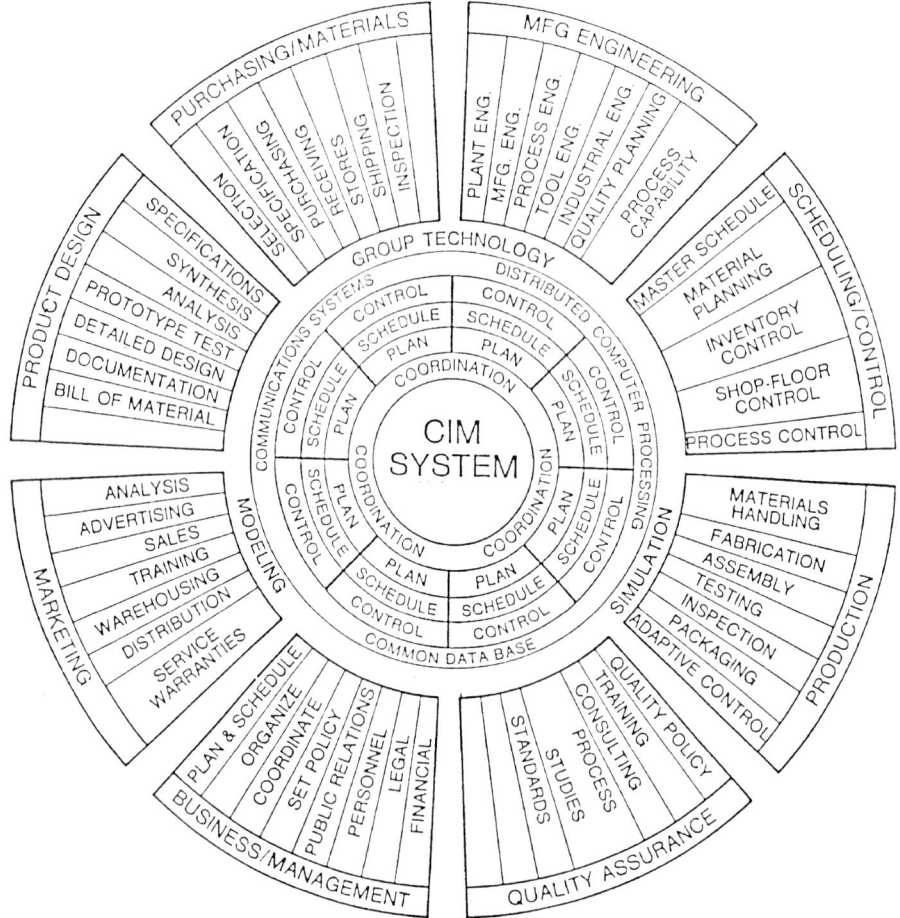

Figure 2.2 CIM system

Figure 2.3 illustrates a manufacturing infrastructure in which the central hub focuses on the customer [4]. The wheel diagram illustrates how everyone in the organization must work toward a common goal for the manufacturing enterprise to be successful. The wheel is much like the Knights of the Round Table, with no dominant group or individual, that is, "all for one and one for all."

2.2.1 Six Perspectives on the New Manufacturing Enterprise

The new Manufacturing Enterprise Wheel describes six fundamental elements for competitive manufacturing [4]:

2.2 Main Structure   15

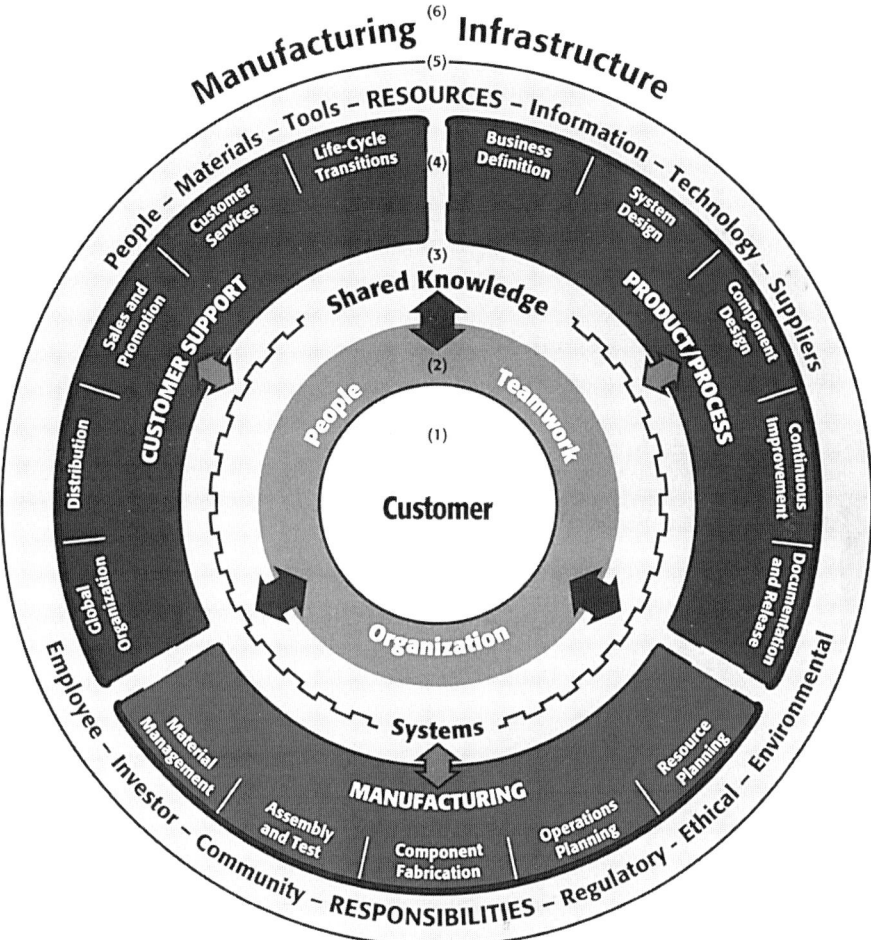

Figure 2.3 Enterprise wheel. CASA/SME Copyright, 1993

1. The central role of the customer and evolving customer needs. A clear understanding of the marketplace and customer desires is the key to success. Marketing, design, manufacturing, and support must be aligned to meet customers' needs. This is the bull's eye, the hub of the wheel, and the vision and mission of the enterprise.
2. The role of the people and teamwork in the organization. Included here are the means of organizing, hiring, training, motivating, measuring, and communicating to ensure teamwork and cooperation. This side of the enterprise is captured in ideas such as self-directed teams, groups of teams, the learning organization, leadership, metrics, rewards, quality circles, and corporate culture.
3. The revolutionary impact of shared knowledge and systems to support people and the process. Included here are both manual and computer tools to aid research, analysis, documentation, decision-making, and control of every process in the enterprise.

4. Key process from product definition through manufacturing and customer support. There are three main categories of processes: product/process definition, manufacturing, and customer support. Within these categories 15 key processes complete the product life cycle.
5. Enterprise resources (inputs) and responsibilities (outputs). Resources include capital, people, materials, management, information, technology, and suppliers. Reciprocal responsibilities include employee, investor, and community relations, as well as regulatory, ethical, and environmental obligations. In the new manufacturing enterprise, administrative functions are a thin layer around the periphery. They bring new resources into the enterprise and sustain key processes.
6. The manufacturing infrastructure. While the company may see itself as self-contained, its success depends on customers, competitors, suppliers, and other factors in the environment. The manufacturing infrastructure includes: customers and their needs, suppliers, competitors, prospective workers, distributors, natural resources, financial markets, communities, governments, and education and research institutions.

2.3 Product Design and Development Management System (PD2MS)

In dealing with the product process many names are given to various methods of accomplishing the product goal. Experience indicates that there are five phases with milestones for every project. (See Fig. 2.4.)

2.3.1 PD2MS

Phase 0: Idea. Is it Feasible?
   Milestone: Define objectives, commitment of resources

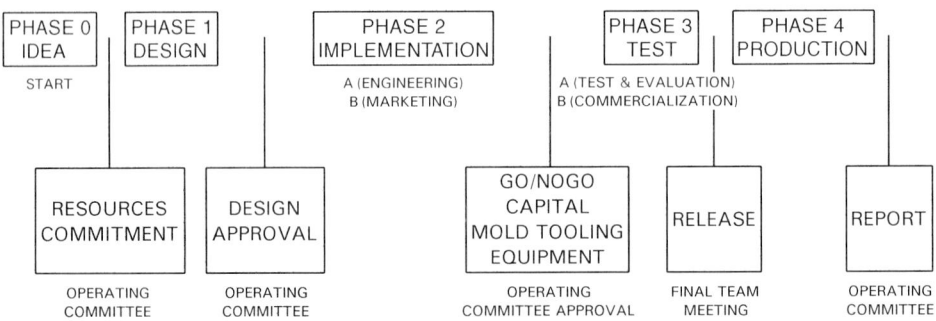

Figure 2.4 Product design and development system

Phase 1: Design. Form and Function
   Milestone: Design approval, meet objectives
Phase 2: Engineering. Is it practical?
   Milestone: Final go/no go, involving capital commitment for machine and tooling
Phase 3: Test, Specify, and Evaluate (revise if necessary)
   Milestone: release of products—meet objectives
Phase 4: Production Run—Feedback
   Milestone: First run report—did we meet objectives?

It is important that written objectives, including the customer's, "must haves," "wants," and "would like to have" are recorded and understood. The designer's role is to have a clear understanding of these criteria, together with the resources, the time schedule, and cost parameters. These also need to be clearly agreed on by the team.

The Team Leader has the additional responsibility to monitor the progress of the project and report to the project owner. The project owner ensures the resources are committed and team cooperation take place.

2.3.2 Process Management Tracking Systems

There are many project management system approaches to programming, scheduling, and contracting that are beyond the scope of this book, including PERT (Program Evaluation and Review Technique), "Precedence" diagramming, and CPM (Critical Path Method). If your organization does not use one, it is advisable to implement your own.

2.3.3 Commitment of Resources

Two basic issues complicate management of resources:

Multiprojects. When several projects are running concurrently, some small, some large, conflicts of priorities and resources occur. Solution: Communication and regular review meetings.

"CHANGES." These always cause problems, but are inevitable. We are on a moving train and nothing remains the same! Solution: Constant review of objectives. Document changes and the effect on schedule and cost.

At first glance, Fig. 2.4 would suggest that the project responsibility is passed from one group to another, with the new product originating with the marketing group. Actually the idea could have come from anywhere but it must be approved by marketing. Then the idea is passed on to the Industrial Design group, who would make renderings and part drawings and then pass it on to the Engineering group. The Engineering group would determine the process and the materials and analyze structural requirements, relying on theoretical analysis and prototype testing. The idea then moves on to Tooling Engineering, who looks at the part from a tooling design and fabrication aspect, and may decide that the part may be difficult or impossible to mold.

Although the system does work, it is time consuming and filled with pitfalls. Returning to the Manufacturing Structural wheel, all groups should be involved in the beginning of the process. This early involvement not only makes sense from the development efficiency point, but also from the human relations viewpoint. Why not "get the person who is going to make the product involved up front?" The process then requires management effort and time to ensure that the objectives are well defined and met on time.

Keeping track of the activities of the individual groups in bringing a product from conception to a production reality can become very complex and time consuming. Further, bringing a product on line, on schedule, on cost, on forecast, on quality, on profit, is a requirement for a successful manufacturing enterprise in today's competitive world.

Computers have become necessary tools to keep up with these activities, and provide timely analysis of the project status, history, and implications on future plans. These activities are illustrated in Fig. 2.5 (on pp. 20 to 25), which shows the computer integrated manufacturing system, a reiteration of the manufacturing infrastructure in terms of group activities.

2.3.4 Concurrent Engineering

In Fig. 2.4 the "Phases" and "Milestones" are documented and the activities of each participating group are interconnected with Product Development and Engineering systems. Figure 2.5 is a flow chart that shows the interconnection between the different groups in the company. Marketing, Manufacturing, Product Development, and Sales all have an interest in getting the product out on time, within a budget, ensure it earns a profit, and so on. Thus by working together many activities can be accomplished concurrently, saving time and wasted effort.

With the computer software programs that are available today with near instantaneous transfer of data and drawings, Concurrent Engineering, has become increasingly important. To apply the Phase system to Blow Molding, Product Design, and Engineering, Figure 2.5 shows the design sequence and priority of activities and decision points for each phase of development.

2.4 Conclusion

Although one cannot be an expert in all aspects of the design arena one can have an understanding of how the system should work. When designing for plastic blow molded parts, one should take into account esthetics, functional process, capabilities, and the manufacturing constraints. As shown, the design process should be an organized approach and should proceed through phases and milestones and include the input of all involved.

20   The Design Process—An Organized Approach

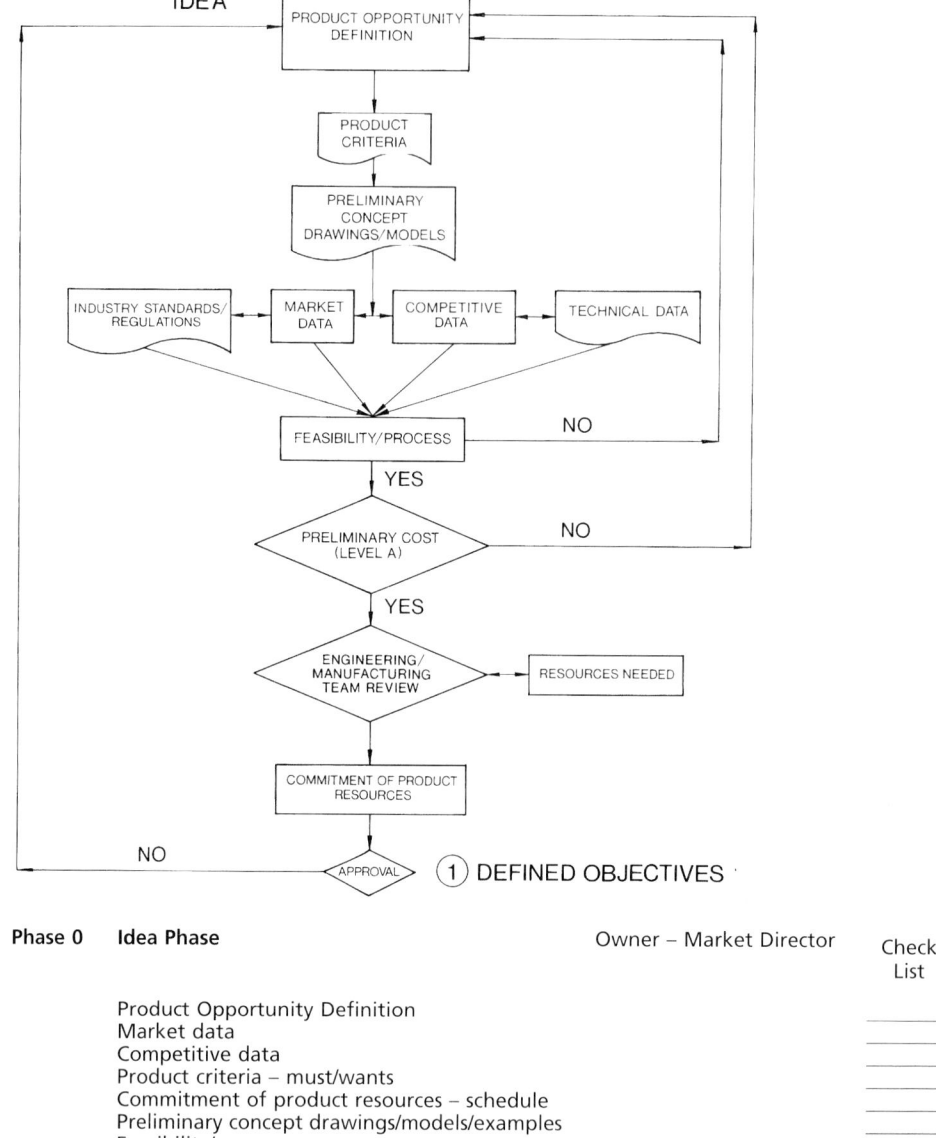

Phase 0   Idea Phase   Owner – Market Director   Check List

　　　　　Product Opportunity Definition
　　　　　Market data
　　　　　Competitive data
　　　　　Product criteria – must/wants
　　　　　Commitment of product resources – schedule
　　　　　Preliminary concept drawings/models/examples
　　　　　Feasibility/process
　　　　　Technical data
　　　　　Industrial standards/regulation
　　　　　Engineering/manufacturing – team review
　　　　　Preliminary cost (level A)
　　　　　Resources needed

　　　　　Operating Committee
　　　　　Approval

Figure 2.5 Activities flow diagram and checklist

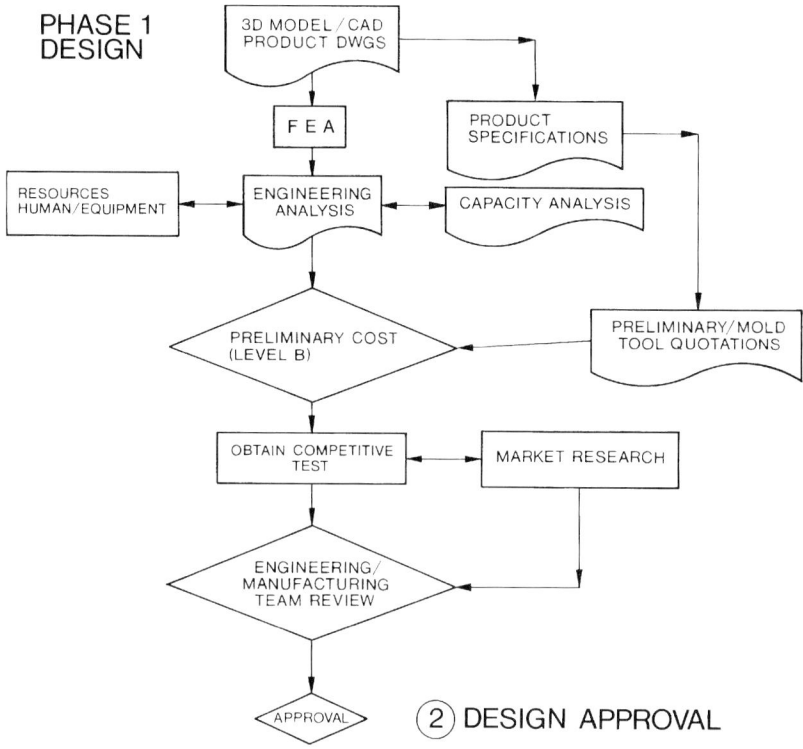

Phase 1	Design Phase	Owner – Design Engineer	Check List
	3D Model CAD		
	Product Drawings		
	Engineering analysis		
	Product specifications		
	Preliminary mold/tool quotations		
	Capacity analysis		
	Engineering/manufacturing – team review		
	Preliminary cost (level B)		
	Obtain competitive product/test		
	Market research		
	Resources – human/equipment		
	Operating Committee		
	Approval		

Figure 2.5 (*continued*)

PHASE 2
IMPLEMENTATION

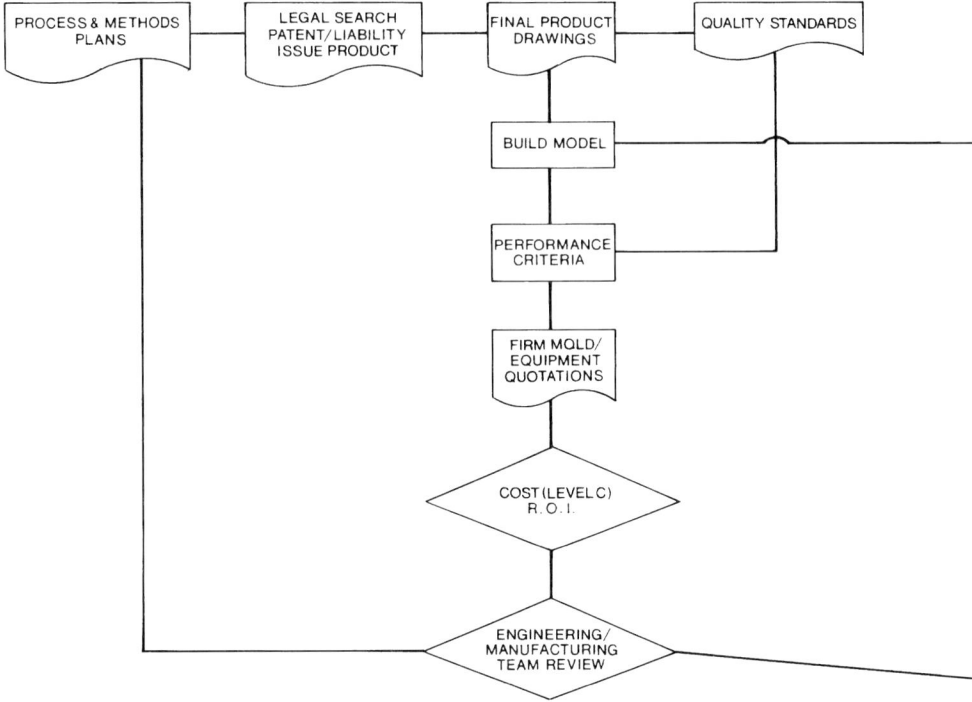

Phase 2A	**Implementing Phase (Engineering)**	Owner – Product Engineer	Check List
	Final product drawings		____
	Build model		____
	Legal search – patent/liability		____
	Issue product performance criteria		____
	Develop initial quality standards – Q.C. plan		____
	Firm quotations – molds & equipment		____
	Firm quotations – purchased parts		____
	Process & methods plans		____
	Engineering/manufacturing review – model/performance		____
	Cost (level C) R.O.I.		____
Phase 2B	**Implementation (Marketing)**	Owner – Marketing Director	
	Market evaluation with model		____
	Promotion plan		____
	Brochures/literature plan		____
	Market data update		____
	Cost estimates		____

Figure 2.5 (*continued*)

```
                    ┌──────────────────┐
                    │ MARKET EVALUATION│
                    │     W/MODEL      │
                    └──────────────────┘
                             │
    ┌──────────────────┐  ┌──────────────┐  ┌──────────────┐
    │BROCHURES/LITERATURE├─┤PROMOTION PLAN├──┤ MARKET DATA  │
    │       PLAN         │ │              │  │    UPDATE    │
    └──────────────────┘  └──────────────┘  └──────────────┘
                             │
                    ┌──────────────────┐
                    │  COST ESTIMATES  │
                    └──────────────────┘
                             │
                    ┌──────────────────┐
                    │    FORECAST      │
                    └──────────────────┘
                             │
                    ┌──────────────────┐
                    │  INTRODUCTION    │
                    │  DATA & PLANS    │
                    └──────────────────┘
                             │
                    ◇   APPROVAL   ◇
                             │
                    ◇ CAPITAL ALLOCATION ◇
                             │
                    ┌──────────────────┐
                    │  CONFIRM SCHEDULE│
                    └──────────────────┘
                    ┌─────────┐ ┌─────────┐
                    │ORDER MOLDS│ │ORDER PURCHASE│
                    │EQUIPMENT │ │   PARTS     │
                    │ TOOLING │ │             │
                    └─────────┘ └─────────┘
```

Phase 2B	**Implementation (Marketing)**	Owner – Marketing Director	Check List
	Forecast Introduction date and plans		___
	Operating committee Approval		
	Capital allocation Place molds, tooling and equipment Confirm schedule Order advance purchase parts		___ ___ ___ ___

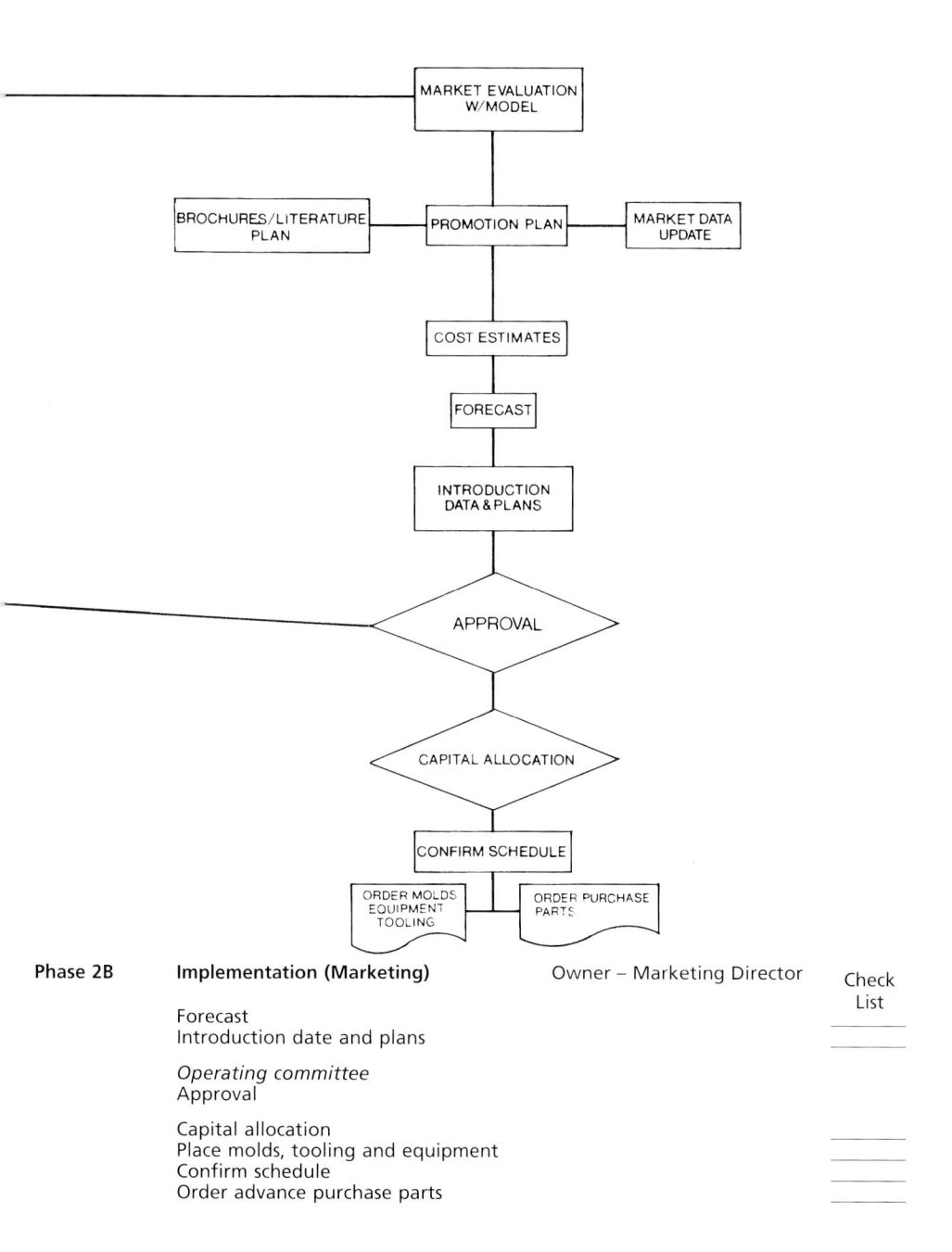

24 The Design Process—An Organized Approach

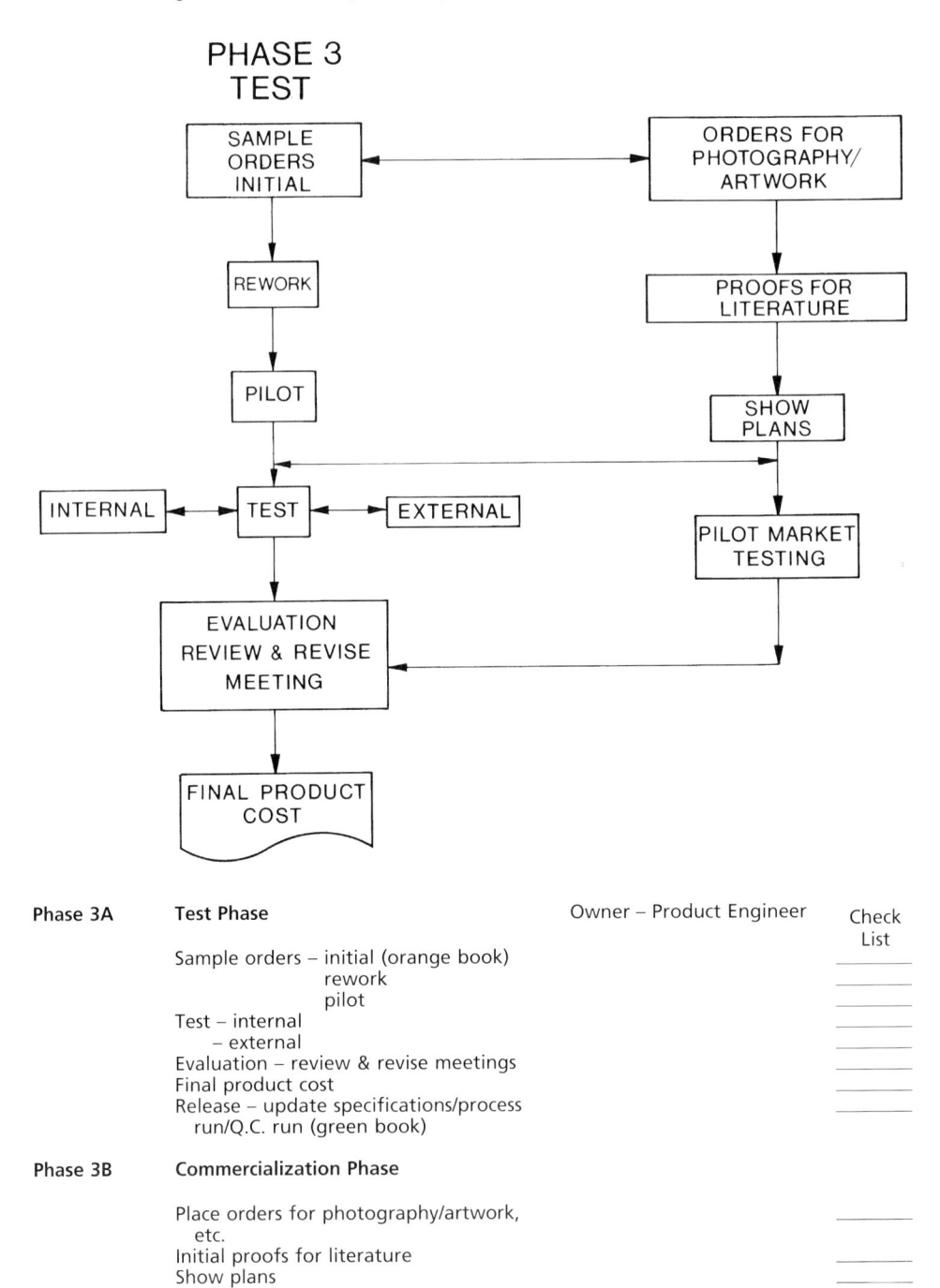

**PHASE 3
TEST**

[Flowchart: SAMPLE ORDERS INITIAL ↔ ORDERS FOR PHOTOGRAPHY/ARTWORK; SAMPLE ORDERS INITIAL → REWORK → PILOT; ORDERS FOR PHOTOGRAPHY/ARTWORK → PROOFS FOR LITERATURE → SHOW PLANS → PILOT MARKET TESTING; PILOT → TEST ← EXTERNAL, INTERNAL ← TEST; TEST → EVALUATION REVIEW & REVISE MEETING ← PILOT MARKET TESTING; EVALUATION REVIEW & REVISE MEETING → FINAL PRODUCT COST]

Phase 3A	**Test Phase**	Owner – Product Engineer	Check List

	Sample orders – initial (orange book)
	    rework
	    pilot
	Test – internal
	   – external
	Evaluation – review & revise meetings
	Final product cost
	Release – update specifications/process run/Q.C. run (green book)		

Phase 3B	**Commercialization Phase**		
	Place orders for photography/artwork, etc. Initial proofs for literature Show plans Pilot market testing		

Figure 2.5 (*continued*)

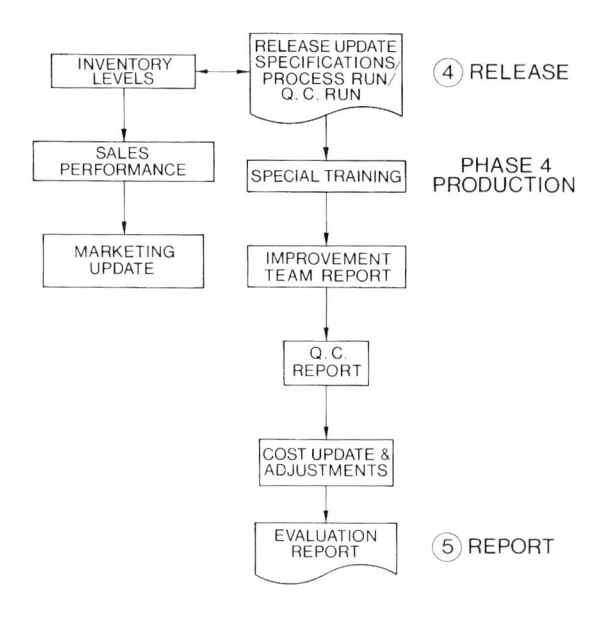

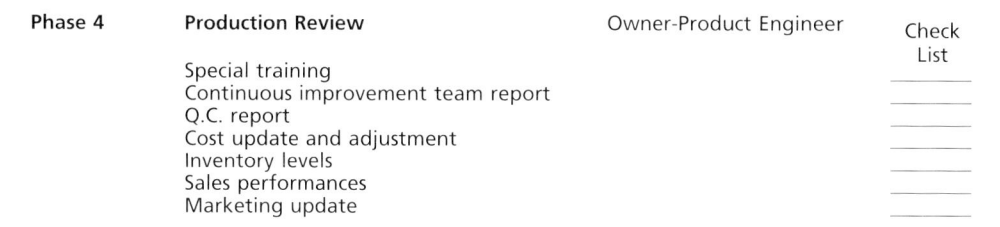

```
                RELEASE UPDATE
                SPECIFICATIONS/
INVENTORY ◄──── PROCESS RUN/      ④ RELEASE
LEVELS          Q.C. RUN
    │                │
    ▼                │
SALES                │              PHASE 4
PERFORMANCE     SPECIAL TRAINING   PRODUCTION
    │                │
    ▼                ▼
MARKETING       IMPROVEMENT
UPDATE          TEAM REPORT
                     │
                     ▼
                Q.C.
                REPORT
                     │
                     ▼
                COST UPDATE &
                ADJUSTMENTS
                     │
                     ▼
                EVALUATION        ⑤ REPORT
                REPORT
```

Phase 4	Production Review	Owner-Product Engineer	Check List
	Special training		____
	Continuous improvement team report		____
	Q.C. report		____
	Cost update and adjustment		____
	Inventory levels		____
	Sales performances		____
	Marketing update		____

Figure 2.5 (*continued*)

2.5 References

1. Allen, D.K., Processing alternatives for cost reduction. *J. Bran Ann* CIRP. (1987) 36, p. x
2. Allen, D.K., Architecture for Computer Integrated Manufacturing. *J. Bran Ann* CIRP. (1986) 35, p. x
3. B.Y.U., CIM Wheel Chart. Department of Technology
4. CASA/S.M.E. The New Manufacturing Enterprise Wheel Manufacturing Infrastructure. The Manufacturing Enterprise Wheel was conceived and produced under the auspices of the Computer and Automated Systems Association of SME (CASA/SME), and builds upon a foundation of earlier work. It represents contributions from scores of experts. Peter Marks led the cross-functional team and served as "system designer." Business definition and project support were maintained by George Hess, Fred Michel, Dan Shunk, and Warren Shrensker in successive terms leading the Board of Advisors. SME resources, including the SME library, were instrumental in research. Many others contributed greatly to "component design" and "continuous improvement" processes, including all those above. Also, the members of the CASA/SME Technical Forum, under the additional leadership and personal involvement of Charles Savage, Warren Shrensker, Barbara Fossum, and Vic Muglia. Nancy Mauter, Association Manager, who took charge of the "manufacturing" process, and many others that contributed

26   The Design Process—An Organized Approach

The following may be used for general reference reading:

1. Kerrner, H., Project Management—A Systems Approach to Planning Scheduling and Controlling. Van Nostrand Reinhold, New York.
2. Rosenau, M.D. Jr., Project Management for Engineers, Van Nostrand Reinhold, New York, 1984
3. Steiner, B., Project Management A.M.R. International, Inc., New York, 1974.
4. Phillips D. H., Todd D.G., Project Management Systems, Planalog, Gladwyne, PA., 1970.

3
Basic Blow Molding Part Design

3.1 Basic Design Considerations

All parts are shaped by the process of blowing air into the inside of the parison, thus enlarging the parison to where it fills the mold surface. The mold, therefore, determines the outside diameter (OD) of the part. The difference between the diameter of the parison and the diameter of the finished part determines to a degree the blow ratio. This would be absolutely true if the parison were blown into a cylinder shape. When the shape is irregular, the basic blow ratio is determined by examining cross-sections of irregular shapes to determine additional blow ratios that occur in these areas. In other words, there are areas of isolated blow ratios in contrast to the general overall blow ratio (The blow ratio is described later.)

3.1.1 Size Variations

When blow ratio variations are taken into consideration, the size of the parison may increase to assist in blowing a uniform part (see Fig. 3.1). In extrusion blow molding, there is the advantage of choosing the size of the parison in relation to the part.

There are times when the blow ratio is acceptable when the parison is encapsulated inside the mold and blown out to the mold side wall. At other times, it is best to have a parison larger then the mold itself and, therefore, the mold will pinch off a part of that parison that will be blown into the mold configuration (see Fig. 3.2).

This positioning of the parison, in relation to the completed part, is essential for efficient molding and to ensure the most uniform wall thickness.

3.1.2 Understanding Hollow Structures

When designing for blow molding, a different thought process is needed than that for single-walled parts, such as those made from injection molding. Understanding the advantages and disadvantages of the blow molding hollow structure and how the configuration can best be utilized are the first steps to a successful part. The designer, when looking at a part being considered for the blow molding process, needs to come to terms with hollow double-walled structures. Designs for most manufacturing processes are

28 Basic Blow Molding Part Design

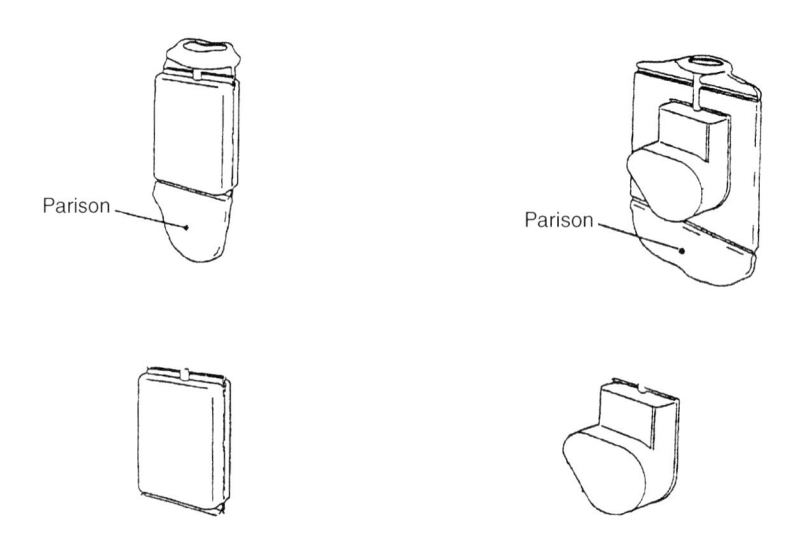

Parison

Parison

Figure 3.1 Uniform part

Figure 3.2 Part larger than cavity

thought of in terms of single-walled, formed or molded into the desired configuration. The choice of wall thickness is governed by the process practicality, appearance, and economic considerations.

Thermoforming, be it pressure or vacuum, is probably the closest comparison to blow molding because it often uses a plastic sheet fit into a female cavity, and two trimmed parts are welded together to provide a one-piece hollow part. Another example would be taking methods used in sheet metal forming and assembly where the strength and stiffness are created by bending, forming, and attaching together by welding or a "pinched" formed seam (see Fig. 3.3).

Those involved in the design process, be it the design engineer, tooling, or process engineer should consider where the mold separates (partline), allowing for draft of the part, layout, blow ratio, corners, finishing, and appearance. The terms listed are defined in the following list.

1. Draft of part. All surfaces parallel to the direction of movement of the mold should have some taper or draft to allow for opening of the mold and easy part removal.
2. The layout of the part as a hollow structure where the outer surface is the part.
3. Look at the total of all functions to be done in the application. Combine those that may be made into one part. A smaller number of parts leads to fewer assembly operations, and therefore reduces cost.
4. Blow ratio. Look at geometry that causes excessive thinning during inflation in the blowing operation. These may be projections or depressions. Excessive width to depth ratios sometimes can be achieved with vacuum assist.
5. Guidelines for radii. Avoid sharp edges and corners, abrupt changes or transitions on all surfaces, and abrupt changes in part diameter or wall thickness.

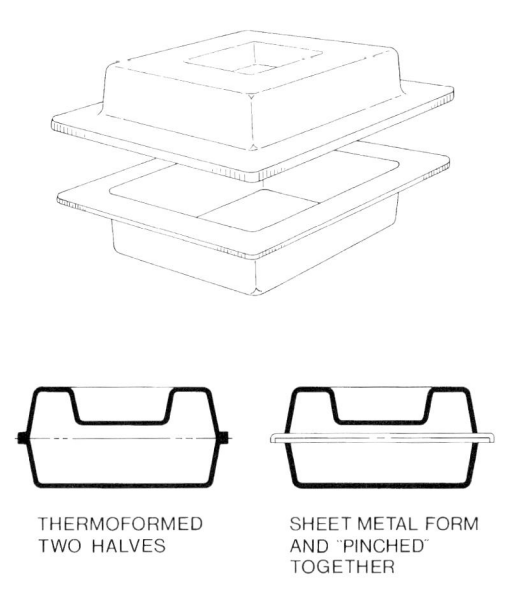

Figure 3.3 Single-shell illustrations

6. Consider giving greater strength and stiffness to the structures, by molding them in geometric configurations such as gussets, ribs, and grooves as alternatives to flat surfaces.
7. Structural parts may use welded surfaces, sometimes referred to as pinch-offs or tack-offs. These are essentially local compression molded areas in a blow molded part.
8. Placement of inserts, openings, vent pins, knock-out holes, lettering, and labels must be carefully taken into consideration.

3.1.3 Draft of Part

Selection of part line and subsequently the draft is the first important step. In most parts, when the part line is flat it is obvious, and the draft a simple matter to add (see Fig. 3.4.a to c).

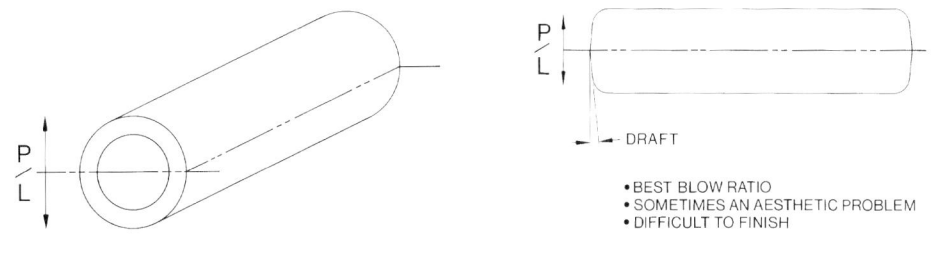

Figure 3.4(a) Simple partline

Figure 3.4(b) Part line location

30 Basic Blow Molding Part Design

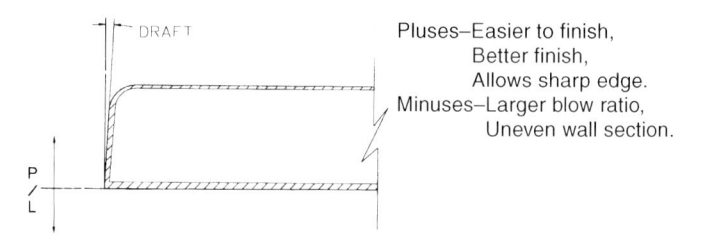

Pluses–Easier to finish,
Better finish,
Allows sharp edge.
Minuses–Larger blow ratio,
Uneven wall section.

Figure 3.4(c) Edge part line

When the part has an irregular part line, figuring the part line becomes more complex. There are some considerations to be given to the mold construction so that no "feather edge" is present (See Fig. 3.5).

The following are general rules for draft:

Minimum draft:
 1 degree/side
Recommended draft:
 2 degree/side
With texture:
 1 degree additional/0.0254 mm depth
 (0.001 in.)

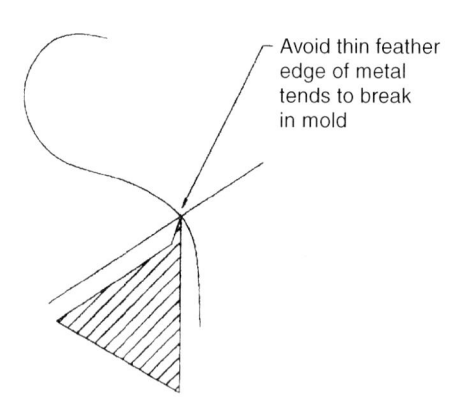

Avoid thin feather edge of metal tends to break in mold

Figure 3.5 Irregular part line

3.2 Increase Draft as Blow Ratio Increases

It should be noted that when a part cools it shrinks from the cavity wall. However, if a core is within the female cavity it will shrink onto the core (see Fig. 3.6); thus the draft angle

3.2 Increase Draft as Blow Ratio Increases

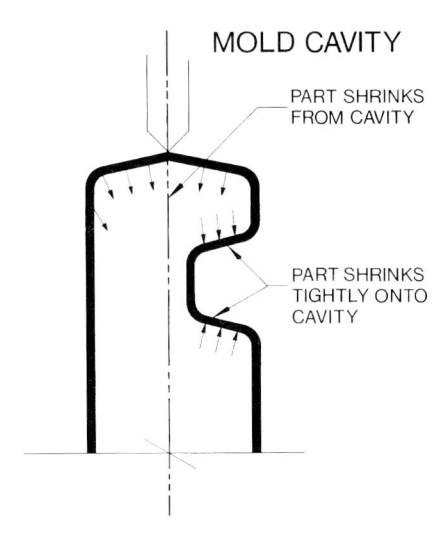

Figure 3.6 Part-shrinkage direction

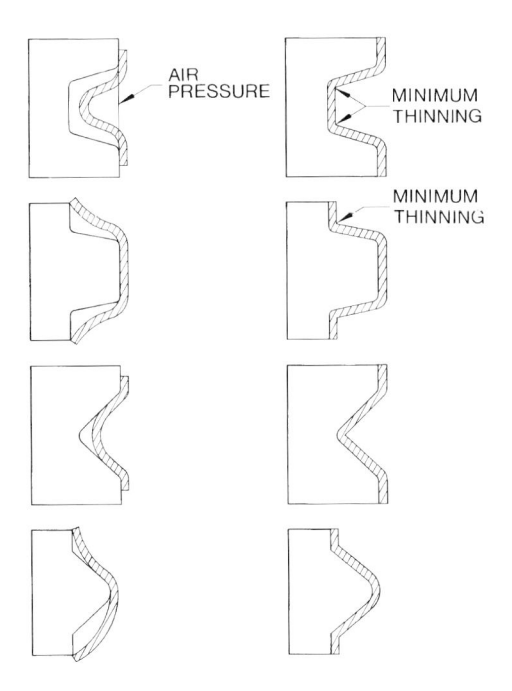

Figure 3.7 Parison inflation

becomes more critical in order to remove the part. The mechanics of inflation are illustrated in Fig. 3.7.

Blow ratio is perhaps one of the most important considerations of the design during the blowing process, where the inflation of the surface of the parison into a cavity has a greater

32 Basic Blow Molding Part Design

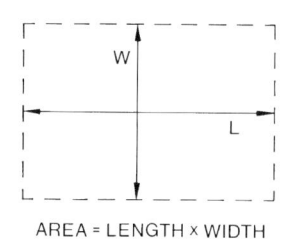

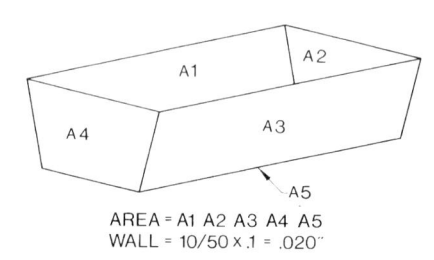

AREA = LENGTH × WIDTH

AREA = A1 A2 A3 A4 A5
WALL = 10/50 × .1 = .020"

Figure 3.8 Theoretical blow ratio—ignoring stretch

surface area and the total volume of material in the parison is the same; therefore as the surface increases the wall thickness must decrease.

Referring to Fig. 3.8, if the area (L × W) is 10 square units and stretches into a cavity 50 square units in surface area and if the initial wall thickness was 0.1 then the average formed thickness is 0.020. This simplistic illusion does not represent the real world, since the actual part does not stretch evenly and thus the final wall thickness depends on geometric conditions of the part.

The factor that is most critical is the width and length dimensions relative to the depth. As a general rule, the depth should not exceed the width or length by more than twofold, when a minimum of 2 degrees draft is used, and a radii of twice the wall thickness.

Thus: Blow ratio $= \dfrac{H}{W}$

$$= \dfrac{\text{(maximum depth of cavity or height of protrusion } V \text{ depth cavity)}}{\text{(smallest dimension at opening of cavity or smallest dimension at base protrusion)}}.$$

A consideration that would allow a greater depth would be a generous radii, top and bottom, and the draft angle (see Fig. 3.9). Another factor to consider would be the shape of the opening (see Fig. 3.10). Four case applications of blow radii guidelines are shown in Fig. 3.11.

Note: A blow ratio optimum may be influenced by materials and the process used. Injection blow molding polyethyleneterephthalate (PET) will have a different optimum than extrusion blow molding high density polyethylene (HDPE).

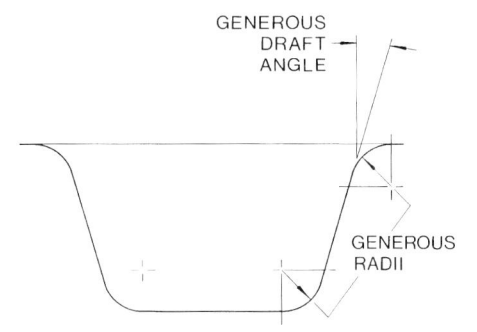

Figure 3.9 Generous draft angle and radii

$$\text{Blow ratio} = \frac{\text{Maximum depth of cavity or height of projection}}{\text{Smallest dimension at opening of cavity or smallest dimension at base of projection}} \quad \frac{H}{D}$$

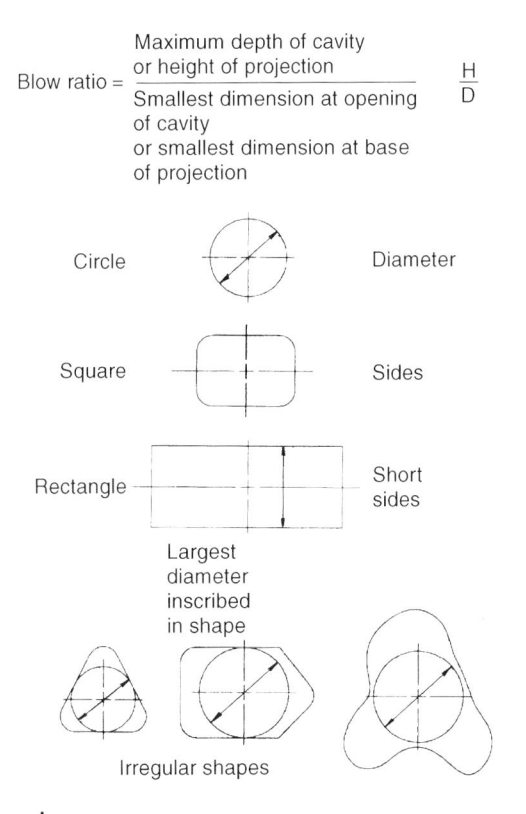

Circle — Diameter

Square — Sides

Rectangle — Short sides

Largest diameter inscribed in shape

Irregular shapes

Figure 3.10 Shape of opening

3.3 Guidelines for Radii

It is generally not practical to achieve very sharp edges or corners without having a heavy wall part to obtain a part that will not blow out in the corner. The following are reasons for specifying generous radii:

General Guidelines for Determining Blow Ratio (BR)

Case 1. Axisymmetric Projection (Female Mold Cavity)

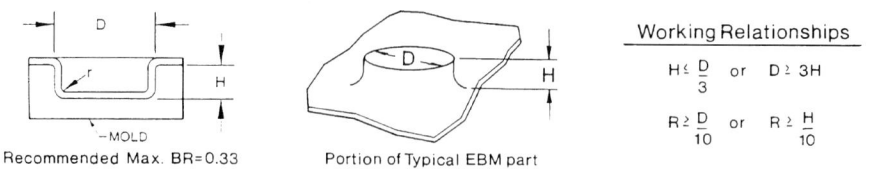

Recommended Max. BR = 0.33

Portion of Typical EBM part

Working Relationships

$H \leq \dfrac{D}{3}$ or $D \geq 3H$

$R \geq \dfrac{D}{10}$ or $R \geq \dfrac{H}{10}$

Case 2. Projection with Plane-Strain* Stretching (Female Mold – Long Channel)

*Plane-Strain means that all stretching occurs in a plane; i.e. is two dimensional.

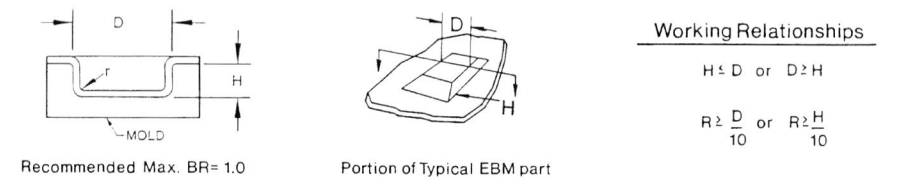

Recommended Max. BR = 1.0

Portion of Typical EBM part

Working Relationships

$H \leq D$ or $D \geq H$

$R \geq \dfrac{D}{10}$ or $R \geq \dfrac{H}{10}$

Case 3. Axisymmetric Depression (Male Pin in Mold)

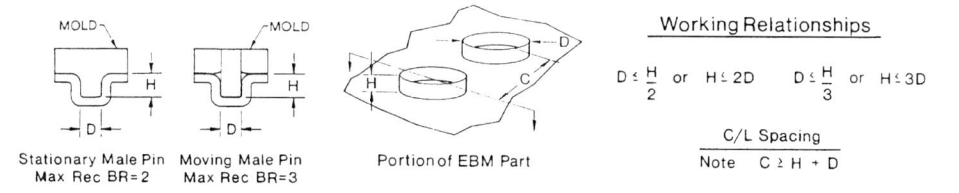

Stationary Male Pin Max Rec BR = 2

Moving Male Pin Max Rec BR = 3

Portion of EBM Part

Working Relationships

$D \leq \dfrac{H}{2}$ or $H \leq 2D$ $D \leq \dfrac{H}{3}$ or $H \leq 3D$

C/L Spacing

Note $C \geq H + D$

Case 4. Depression with Plane-Strain Stretching (Long Male Projections in Mold)

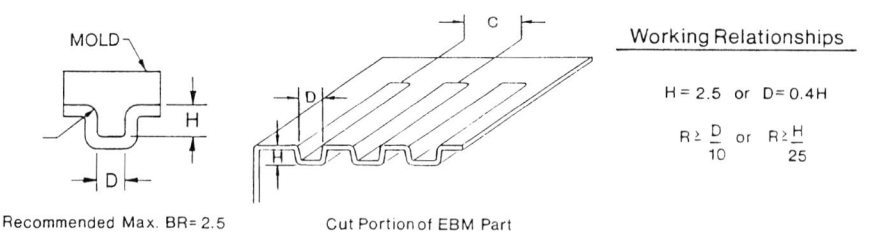

Recommended Max. BR = 2.5

Cut Portion of EBM Part

Working Relationships

$H = 2.5$ or $D = 0.4H$

$R \geq \dfrac{D}{10}$ or $R \geq \dfrac{H}{25}$

Figure 3.11 General guidelines for determining blow ratio

1. The material will thin excessively and also blowout by not filling the corner or edge.
2. When a part does not blow properly, a sharp corner or edge will cause a significant increase of stress, leading to bending, fatigue, and impact strength problems in the final molded part.

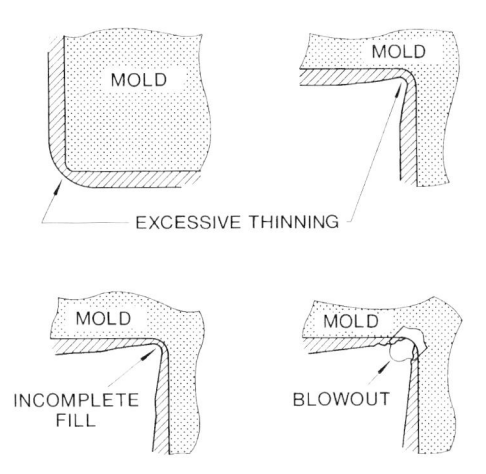

Figure 3.12 Inside/outside radii

3. Thinning will occur at shape edges and corners in the cavity.
4. Sharp corners or a small projection inside the mold cavity limits the achievable blow ratio.

As illustrated in Fig. 3.12, parts may be classified into two groups: those with inside radii, and those with outside radii. Outside radii occur when the parison inflates into the mold whereas inside radii occur when a projection on the mold pushes a cavity or depression into the part. Outside radii are usually more critical since material stretching occurs when the parison inflates into the mold cavity.

3.3.1 Corner and Edge Rounding [1]

To maintain an even wall thickness in blown articles, it must be assumed when designing the part and mold that the sections of the parison resting against the mold wall when blowing commences will be stretched, causing excessive thinning and even blowout. Corners and edge that are formed last must therefore be suitably rounded off. For cylindrical molding, the radius of the edge rounding should not be less than 1/10 of the container diameter; for parts with an oval cross-section this applies to the smallest diameter. The minimum permissible value for corner rounding on square molding can readily be determined graphically as shown in Fig. 3.13. According to this, the radius for corner rounding is:

$$r_E \geq \frac{tF}{2}(1 - \sin 45°) \geq 0.15 tF$$

However, it should if possible be slightly larger because of the multiaxial stretching.

To avoid a notch effect, all edges on threads, ribs, corrugations, and ornamental strips should be rounded off. If especially high shock resistance is required, for example for thin-

36   Basic Blow Molding Part Design

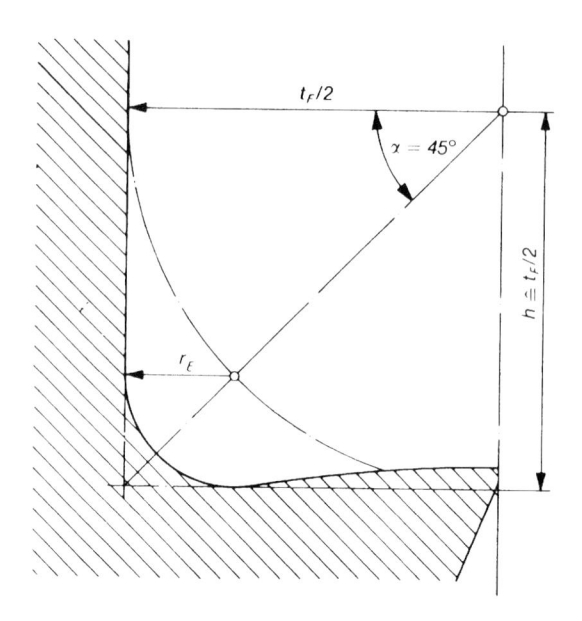

Figure 3.13 Graph for determining the corner rounding of square molding. t_F depth of mold cavity, h reference dimension, r_E radius for corner rounding (minimum permissible value) (Courtesy of Hoechst A.G.)

walled disposable packs, fancy designs must be abandoned. No ridge will appear along the mold parting line if both mold halves are carefully aligned. Inserts, such as base inserts in bottle molds, should not be used. These design principles apply also to heavy duty containers.

3.3.2 Chamfers

Chamfers are often a better alternative to radii in blow molding parts because they lend a better appearance in the design. They give a high tech and square look and stretch the polymer less (see Fig. 3.14).

3.4 Molded-In Geometric Configuration

The double-walled hollow structure of the blow molded process is exceptionally well suited to the inclusion of geometry that aids in rigidity, strength, and fastening points (see Fig. 3.15).

Ribs and gussets are very effective ways to increasing strength and stiffness in a structure (see Fig. 3.16).

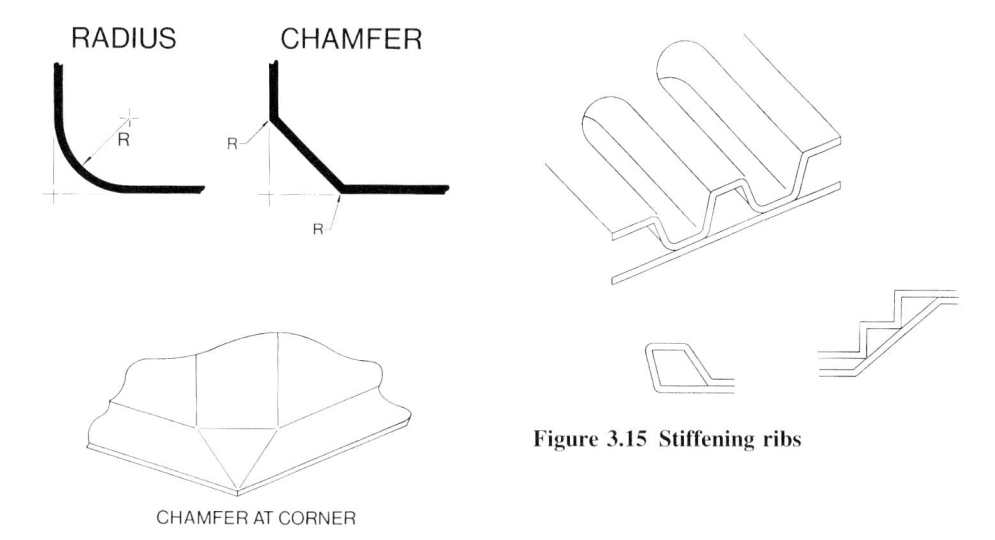

Figure 3.14 Radii vs chamfer

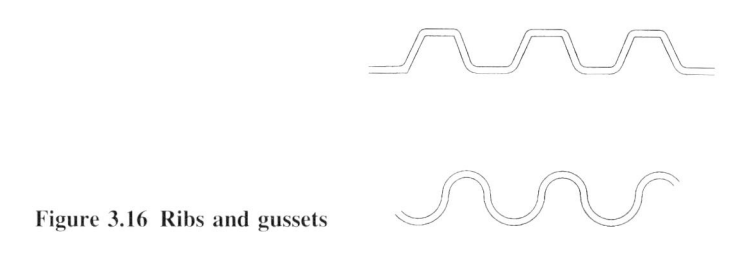

Figure 3.15 Stiffening ribs

Figure 3.16 Ribs and gussets

Generally speaking flat surfaces should be avoided, as there is a tendency for them to flex or "oil can," a tendency of thin-walled parts. Alternatives therefore should be found (see Fig. 3.17).

3.5 Flanges and Tack-Offs

Many other design considerations must be observed to ensure maximum efficiency of the design. Blow molded parts are hollow and, therefore, make a complete unit. In large panel areas which are relatively thin in cross-section, it may be necessary to corrugate or tack-off one side of the panel to provide stiffness and/or support. At other times using flanges may provide the necessary results to make the part rigid.

Flanges and tack-offs are compression molded, in which both sides of the parison are combined, and they should be $1\frac{1}{2}$ to 2 times the parison thickness. Do not compress beyond

Figure 3.17 Dome and arch

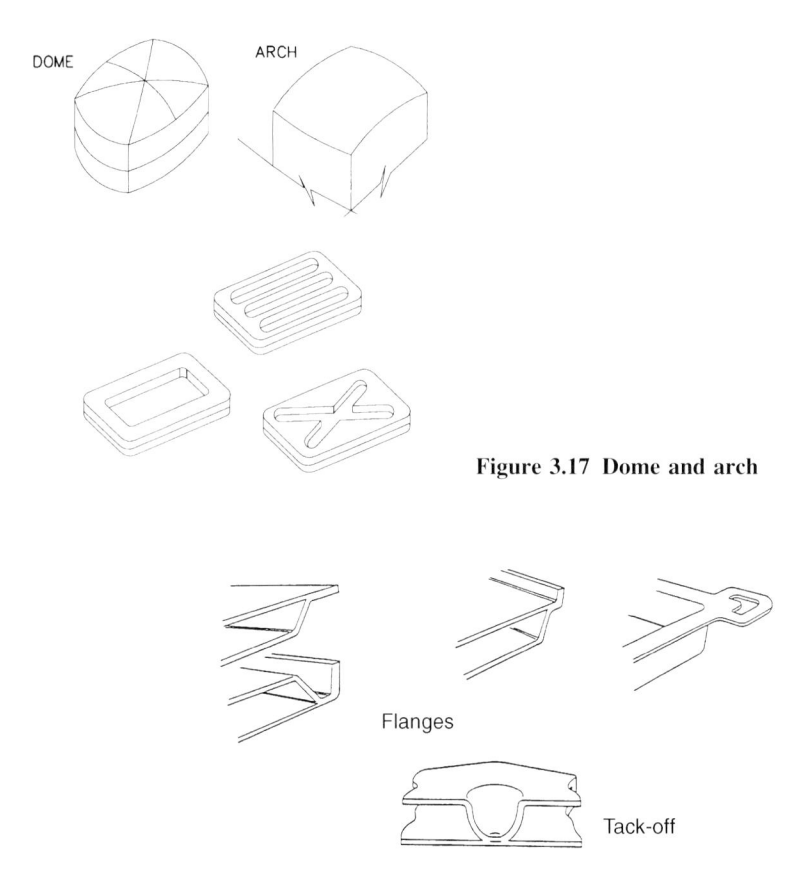

Figure 3.18 Flanges and tack-offs

this otherwise, the part will show sinks or surface blush as well as tearing if the design is too sharp (see Fig. 3.18).

Holes can be molded into the part wall and into flange areas. Core pins can provide sections for screws, expansion inserts, and self-threading inserts, along with a variety of adhesive and solvent bonding inserts.

3.5.1 Threaded Parts

Threaded parts can be blow molded both externally and internally. Usually, external threads are located on the parting line (Fig. 3.19). Internal threads can be an insert (such as a injection molded part) that is placed on the blow pin and encapsulated on the parison and become part of the finished product. Another method is to have a removable blow tip that has the machine thread and is unscrewed at the end of the blow cycle.

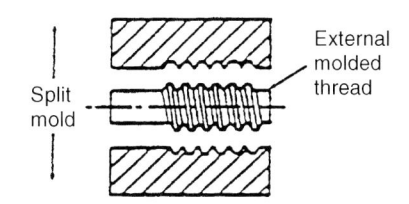

Figure 3.19 External thread

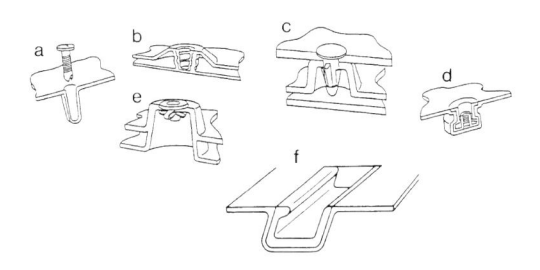

Figure 3.20 Metal and plastic inserts, (a) Threaded metal insert mold in during part formation or ultrasonically inserted into molded recess after part formation. (b) Molded-in plastic part, such as inspection-molded handle

3.5.2 Attachments and Auxiliary Units

In addition, other attachments and auxiliary units can become an integral part of the blow molded part by blowing the parison over components which have been inserted in the mold. The mold insert requires a design that allows the parison to flow around it, capturing it and thus making it part of the finished product. These inserts may be metal or many times are injection molded (see Fig. 3.20). Figure 3.21 shows an application of inserts in computer office furniture and an example of the application of tack-offs and structural ribs.

All of these design considerations must be reviewed in relation to mold design, since these features are incorporated into the mold body or into the pinch-off areas of the mold.

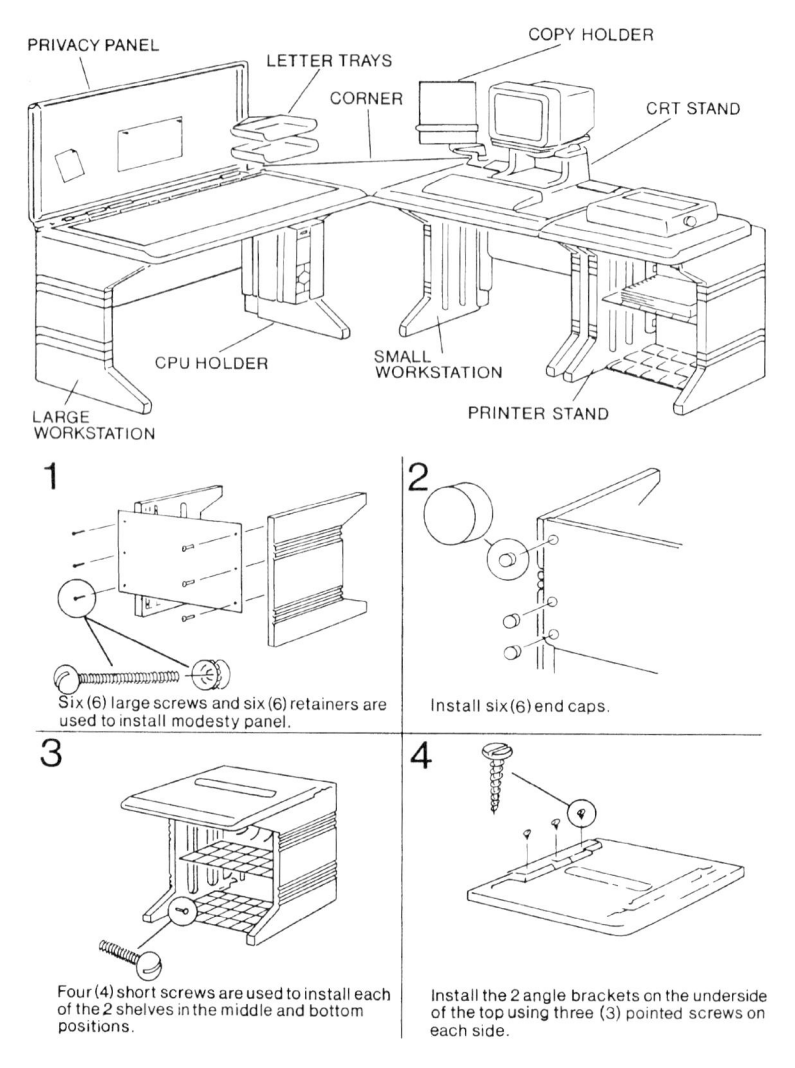

Figure 3.21 Example of threaded inserts and use of structural ribs in office furniture

3.6 Conclusion

A vast amount of blow molding knowledge and technology has evolved over the last decade and an integrated methodology has developed. The designer should consider the double wall configuration of blow molded parts and how this may be applied in such applications as large structural parts, and fluid and kinetic energy management, among others. The designer should be familiar with how the function can be enhanced with ribs, gussets, hinges, and interlock of the dual wall structure. These are covered in the following chapters. Figure 3.22 shows a structural part being molded.

Figure 3.22 Molding structural parts

Acknowledgments

The basis for the work in this chapter is credited to the team from GE Plastics and their study of engineered blow molded plastics.

This effort was captured in the publication by GE of "Engineered Blow Molding Part Design," Author Lincoln J. Alvord.

Although the focus was on engineered plastics the writer has found from experience the same principles and methodology applies to polyolefins.

3.7 Reference

1. Hoechst Plastics, Blow Molding of Plastics, Hoechst Aktiengesellschaft Verkauf Kunststoffe Frankfurt, Germany

4
Design for Bottles

The packaging industry, as any other, must continuously strive for high productivity while maintaining quality assurance to provide optimum value of its products.

A way to reduce the cost of packaging, without sacrificing market requirements, is to use containers made by the blow molding process, with the resin content as the single most important cost element. This is the logical place to begin. Thus minimizing the weight of material in the container will reduce cost. To achieve effectiveness in manufacturing and still meet packaging requirements, a compromise between container weight and machine productivity is usually dictated. Design, aesthetics, consumer conveyance, processing, and container performance are all factors to be included in the equation, to produce a functional cost-effective part.

4.1 Blow Molding Process Basic Shapes

Optimizing a container design from the blow molding process requires consideration of three basic shapes:

1. Circular tube—injection blow molded with a closed bottom test tube shaped parison (Fig. 4.1);
2. "Toothpaste" tube—extrusion blow molded in which the parison falls inside a centrally located neck; ideal for bottles that are oval at the bottom and round at the top (Fig. 4.2); and
3. flat "pillow"—extrusion blow molding where the parison falls outside the neck (Fig. 4.3).

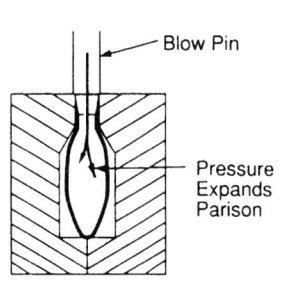

Figure 4.1 Circular tube

44 Design for Bottles

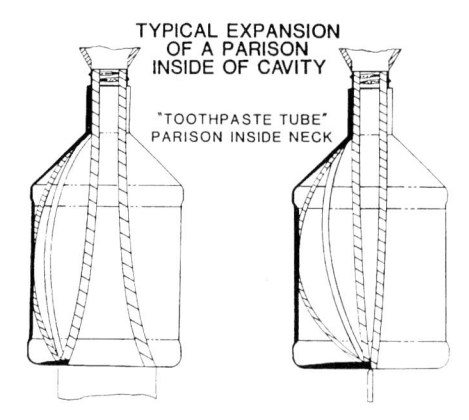

Figure 4.2 "Toothpaste" tube

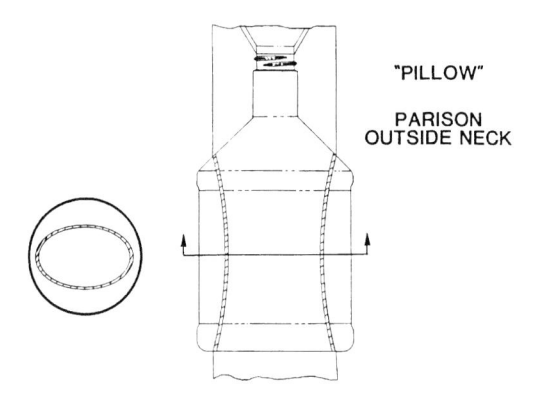

Figure 4.3 Flat pillow

4.2 Simplified Assumptions About Parison Expansion

(The actual mechanics are more complicated.)

The parison cross-sectional shape expands, but before it strikes a mold surface it remains relatively consistent and the wall thickness will thin uniformally.

The parison thickness freezes the instant it strikes the mold surface. The remaining parison still in free space will continue to thin at a uniform rate until it strikes a mold surface.

4.3 Bottle Design Concepts

There are four basic rules for extrusion blow molding of HDPE. (Those for other resins and injection blow molding are slightly different.)

4.3 Bottle Design Concepts

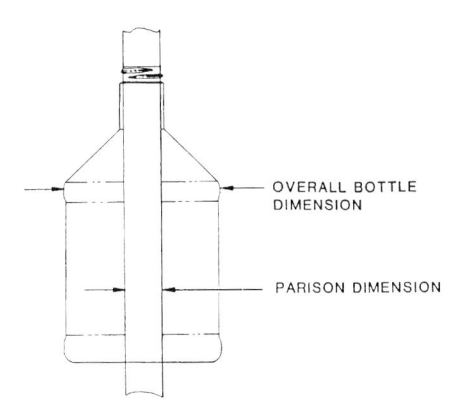

Figure 4.4 Blow-up ratio

1. Blowup ratio of no more than 4 : 1 (Fig. 4.4), measured as parison diameter to minimum bottle diameter. This rule also applies to bottle sections such as the handle. The lower this ratio, the more consistent the wall thickness. A heavy bottom with "high" blowup ratio will add weight and cooling cycle time but little or no strength.
2. Radius, slant, and tapering all surfaces (Fig. 4.5). Square and flat surfaces with sharp corners, although offering cubic shaping efficiency, will not work (Fig. 4.6). Wall thickness will vary considerably with thin corners that are weak, and panels that are thick and distorted. The latter is caused by poor cooling and difference in shrinkage rates. Figure 4.7 shows an improved bottom container contour. Also, flat surfaces offer little top load strength. Figure 4.8 shows an example of shoulder contour and neck ring.
3. Sharp accent lines can cause problems owing to trapped air. This air is caught momentarily between the mold surface and the plastic pressing against it; which causes the plastic to become unusually thin at the line and in turn can cause failures in drop impact performance.
4. Always avoid abrupt changes in cross-sections and in profile.

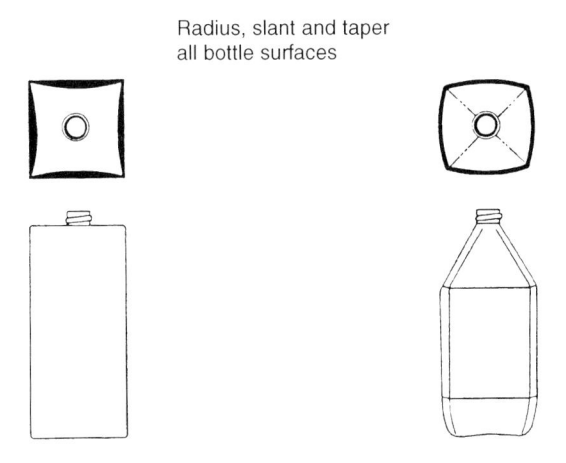

Figure 4.5 Radius slant, and taper—all bottle surfaces

46 Design for Bottles

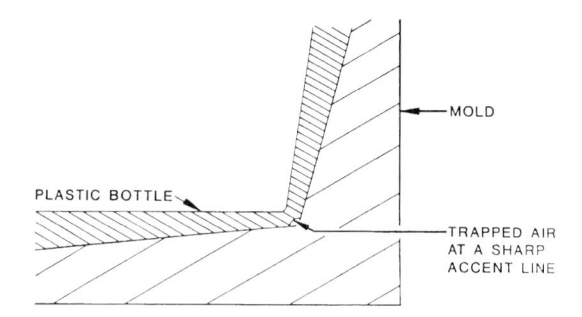

Figure 4.6 Accent line air entrapment

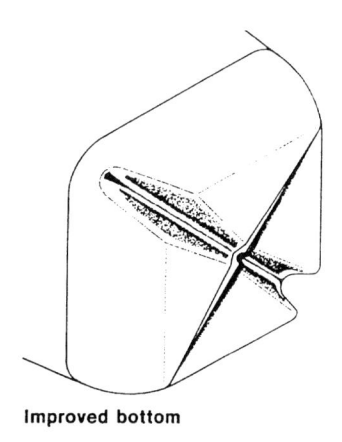

Figure 4.7 Improved bottom contour
(Courtesy of Johnson Controls)

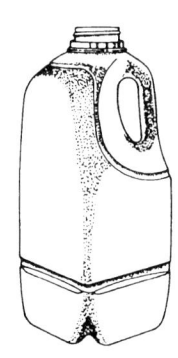

Figure 4.8 Shoulder and neck ring
(Courtesy of Johnson Controls)

4.3.1 Ribs: Do Not Always Stiffen

The wall is to be thinned when more surface area for the parison to cover is created. Horizontal ribs create bellows or an accordion effect which flexes easier. This design can be used effectively if the designer wishes the part to flex; however, this should be avoided if stiffness is required. Analyze the structure, and if flexing is to happen, note where the hinge point will occur. Take steps to alter the design to interrupt the hinge action. Figure 4.9 shows horizontal and vertical ribs.

As a general rule, use a round container to improve hoop stiffness, but pay attention to the cross-section so as not to create the accordion condition mentioned in the previous paragraph. Square containers actually reduce stiffness, thus reducing both top load strength and bulge resistance as well. Figure 4.10 shows a truss groove.

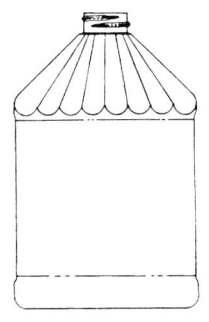

Figure 4.9 Horizontal and vertical ribs

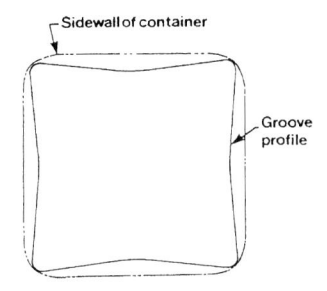

Figure 4.10 Truss groove

4.3.2 Cross-Sections

Establish the partline so that the part has an acceptable blow ratio, and the cavity geometry is not "oversquare" (sections that are more deep than wide) (Fig. 4.11), a problem often seen in the handle and occasionally with engraved lettering (Fig. 4.12).

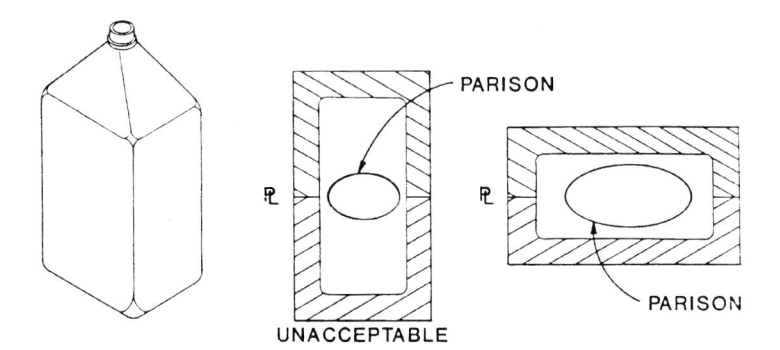

Figure 4.11 Cavity oversquare

48 Design for Bottles

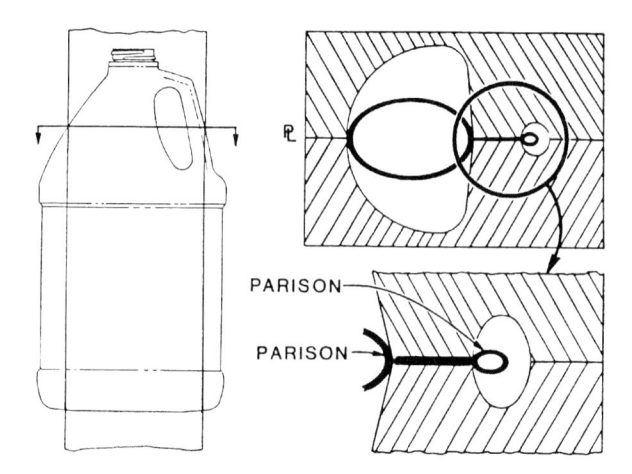

Figure 4.12 Example of an oversquare condition in bottle handle

4.3.3 Bottle Neck, Threads, and Openings

The blowing process generally uses a blow pin with a channel to emit air into the parison, creating a natural opening for bottles and containers. There are exceptions to this method of blowing air, discussed elsewhere. In the neck section, the thickness can be controlled to a greater degree than in the rest of the part with the careful matching of the blow pin and neck cavity dimensions. Actually this area may be treated as compression molding. Of course some type of plug or cap must be used to fill the hole when the container and contents are to be stored or transported. Since the majority of blow molded containers are used for packaging, it is important to have standards for threaded openings, which are to be closed with a cap. Recommended voluntary guidelines for the dimensions have been established by the plastic bottle division of the Society of the Plastic Industry. An example of this is shown in Fig. 4.13. Two main shapes, L and M, are shown in the figure. The L shape is similar to those found on glass bottles, while M is typical of plastic containers. A complete listing can be obtained from the Plastic Bottle Institute [1]. An example of SPI charts is shown in Table 4.1.

Other neck openings are designed for caps or spray attachments. Typical designs are shown for these types of neck configurations in Fig. 4.13 and Fig. 4.14.

4.4 Container Volume

Traditionally dairy containers have been filled to the very top. Consumers expect the plastic dairy containers they use today to be filled in the same way. Dairymen, for the lowest possible cost, demand containers not contain more than the stated volumetric measure. Any

4.4 Container Volume

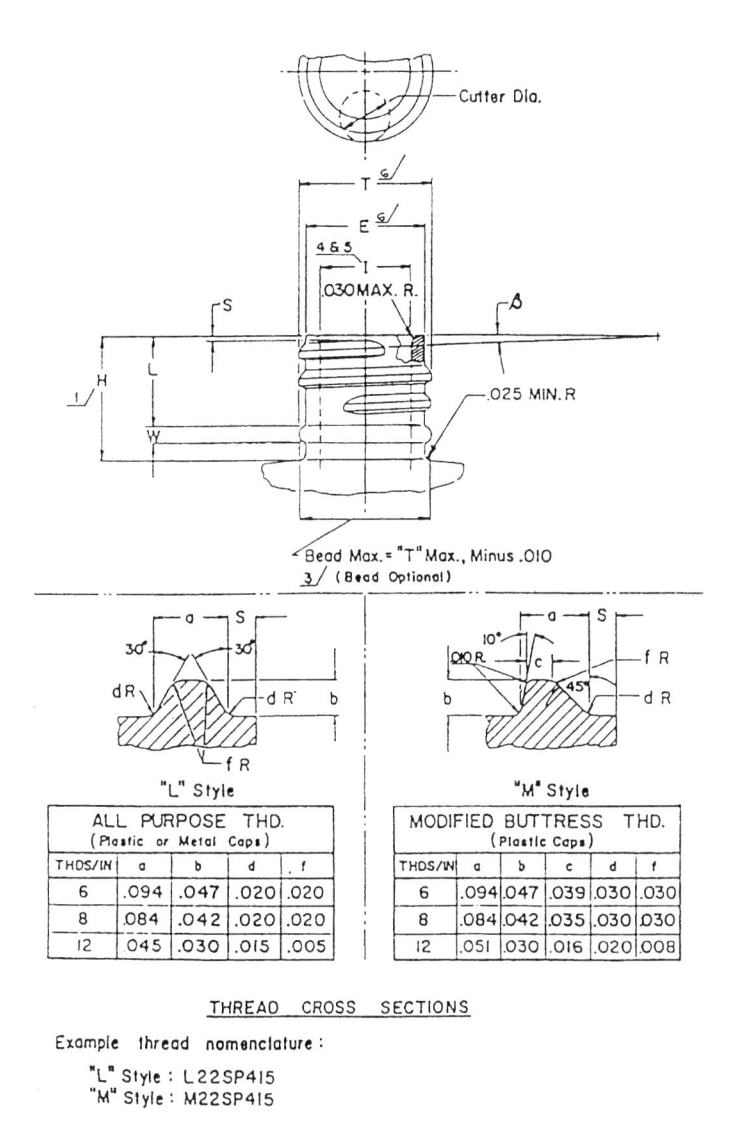

Figure 4.13 Example of SPI thread neck standards

extra milk that is added to fill the container will quickly reduce profits. To achieve this target, the overall volume of plastic dairy containers must be "fine-tuned." The volume of the container is adjusted to the precise running conditions and age at fill. The purpose of this section is to discuss how dairy containers are measured for volume and how molds are corrected at the mold makers; second, to identify sources of error in volume measurements, and third, to identify molding conditions that can easily change container volume.

50 Design for Bottles

Table 4.1

SP-415 FINISH

mm	T[f] Max.	T[f] Min.	E[f] Max.	E[f] Min.	H[a] Max.	H[a] Min.	L Min.	S Max.	S Min.	I[d,e] Min.	W[c] Max.	Helix angle β	Cutter diameter	Threads per inch
13	0.514	0.502	0.454	0.442	0.467	0.437	0.306	0.052	0.022	0.218	0.045	3°11'	0.375	12
15	0.581	0.569	0.521	0.509	0.572	0.542	0.348	0.052	0.022	0.258	0.045	2°48'	0.375	12
18	0.704	0.688	0.620	0.604	0.632	0.602	0.429	0.052	0.022	0.325	0.084	3°30'	0.375	8
20	0.783	0.767	0.699	0.683	0.757	0.727	0.456	0.052	0.022	0.404	0.084	3°7'	0.375	8
22	0.862	0.846	0.778	0.762	0.852	0.822	0.546	0.052	0.022	0.483	0.084	2°49'	0.375	8
24	0.940	0.924	0.856	0.840	0.972	0.942	0.561	0.061	0.031	0.516	0.084	2°34'	0.375	8
28	1.088	1.068	0.994	0.994	1.097	1.067	0.665	0.061	0.031	0.614	0.094	2°57'	0.500	6

SP-410 FINISH

mm	T[f] Max.	T[f] Min.	E[f] Max.	E[f] Min.	H[a] Max.	H[a] Min.	L Min.	S Max.	S Min.	I[d,e] Min.	W[c] Max.	Helix angle β	Cutter diameter	Threads per inch
18	0.704	0.688	0.620	0.604	0.538	0.508	0.361	0.052	0.022	0.325	0.084	3°30'	0.375	8
20	0.783	0.767	0.699	0.683	0.569	0.539	0.361	0.052	0.022	0.404	0.084	3°7'	0.375	8
22	0.862	0.846	0.778	0.762	0.600	0.570	0.376	0.052	0.022	0.483	0.084	2°49'	0.375	8
24	0.940	0.924	0.856	0.840	0.661	0.631	0.437	0.061	0.031	0.516	0.084	2°34'	0.375	8
28	1.088	1.068	0.994	0.974	0.723	0.693	0.421	0.061	0.031	0.614	0.094	2°57'	0.500	6

[a] Dimension "H" is measured from the top of the finish point where diameter T, extended parallel to the centerline, intersects the top of the shoulder.
[b] A MINIMUM OF 1½—full turns of thread shall be maintained.
[c] Use of bead is optional. If bead is used, bead dia. and "L" MINIMUM must be maintained.
[d] Hole dia. "I" to be measured through full length of finish unless otherwise specified.
[e] Concentricity of "I" Min. with respect to diameters "T" and "E" is not included. "I" Min. is specified for filler tube only.
[f] "T" and "E" dimensions are the average of two measurements taken 90° apart. The limits of ovality will be determined by the container supplier and container customer, as necessary.
[g] All dimensions are in inches, unless otherwise indicated. To the best of our knowledge the information contained herein is accurate. However, The Society of the Plastics Industry, Inc., assumes no liability whatsoever for the accuracy or completeness of the information contained herein. Final determination of the suitability of any information or material for the use contemplated, the manner of use and whether there is any infringement of patents is the sole responsibility of the user.

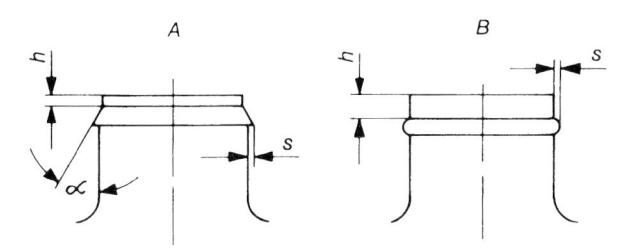

Figure 4.14(a) Pouring neck with click stop for snap closure. A, Design with conical torus. h = 1–3 mm, a = 30–450, s = 1–2 mm. B, Design with annular torus, dimensions as in A, (Courtesy of Hoechst A.G.)

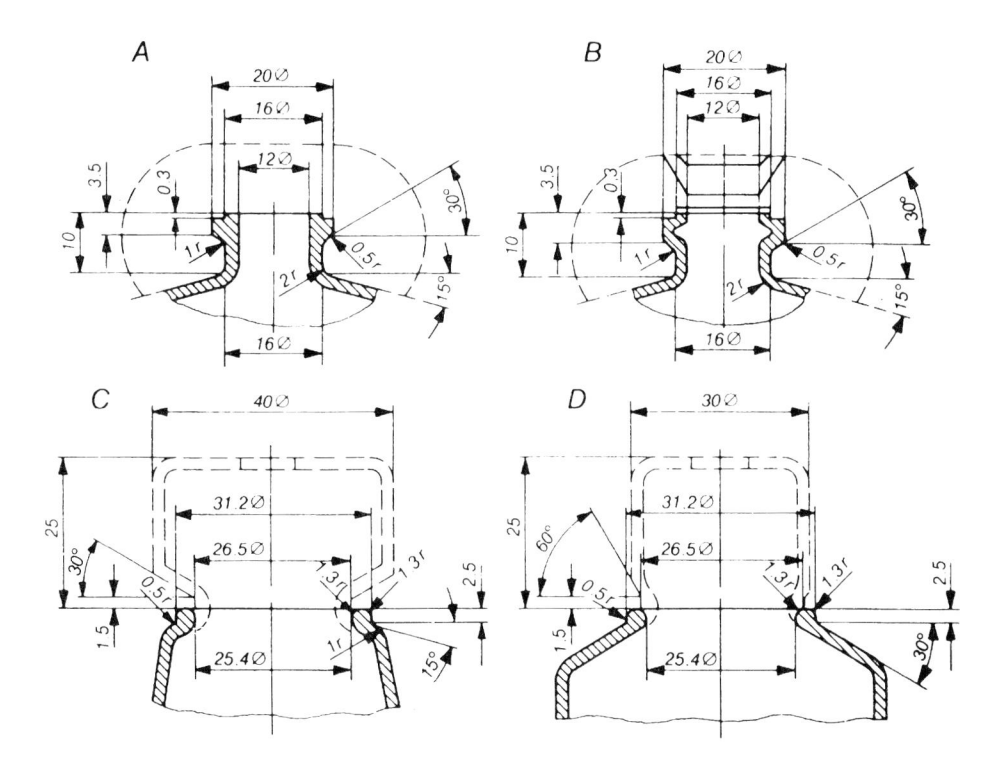

Figure 4.14(b) Design recommendations for the top of an aerosol container. A, In internally calibrated glass bottle type top. B, Externally calibrated glass bottle type top. C, Top for 1 in. lift valve on slimline aerosol container. D, Top of 1 in. lift valve on aerosol shouldered container. (Courtesy of Hoechst A.G.)

4.4.1 Container Volume Measurements

One gallon is not 8.3 pounds of water, or 8.6 pounds of milk, and does not change at warmer or colder temperatures. One gallon is, by definition, 231 cubic inches. Products such as water, milk, and fruit juices expand or contract with changes in temperature. Water, for example, is at its highest density at 4 °C (39.4 °F). At cooler or warmer temperatures water expands. One gallon (231 in.^{3}) of water at 4 °C (39.4 °F) will weigh more than one gallon (231 in.^{3}) of water at 20 °C (68 °F). In other words, 1 gallon of water measured at 20 °C (68 °F) will be approximately 7.3925^{-3} liters (1/4 fl oz.) less than 1 gallon if cooled to 4 °C (39.4 °F).

Weight can, however, be used to determine container volume as long as precise conditions are maintained. One gallon of distilled water at 4 °C (39.4 °F) weighs 3781.082 g.

Every effort should be made to duplicate identical production conditions during the machine line run when sample containers are saved for volume checking. Containers are generally measured "fresh," that is, measured at an average age of one half hour old. The age of the container is important because HDPE resin after molding continues to shrink over a period of time. Storage temperature, which is assumed to be room temperature, controls the rate of shrinkage. When the container if filled with cold milk and stored cooler, the rate of shrinkage is reduced to virtually zero for the remaining expected life of the container. In effect filling freezes the container at whatever size it is at that time. The following procedure should be used to measure machine line, sample containers and to average the effects of age:

- Five samples from each cavity are weighed and grouped.
- The first sample of each cavity is quickly filled to the brink of overflow with distilled water at 4 °C (39.4 °F). The weight of the empty container and the weight of the container with water are recorded.
- Next the second sample of each cavity is checked, then the third and so forth until all containers have been measured.
- Halfway through the measuring of the entire group of samples a second bottle of water is substituted to reduce the error from the slight warming of the water as it is transferred from container to container.
- The average net weight of water is calculated for each cavity and compared to the standard.

4.4.2 Standards

One of two overflow weight standards is used for comparison depending on the intended use of the container inserts. The standards are based on the following (Fig. 4.15):

3781 g	Weight of one gallon of distilled water at 4 °C (39.4 °F) to fill level ($\frac{1}{2}$ in.) from the top of container.
15 g	Weight added for finished head space (see drawing I)
9 g	Weight added for safety factor
3805 g	Overflow weight Standard I

The standard of 3805 g is used under three conditions: first, when containers are measured "aged," a minimum of 48 hours old, and with no volume control inserts or with

4.4 Container Volume

```
.5 FL OZ ─────────────          ───── 15 g TO OVERFLOW*
                                ───── 24 g TO OVERFLOW*
1.25 FL OZ ──────               ───── 34 g TO OVERFLOW*
1.5 FL OZ ────────

                          *Grams of distilled water at 4° C.
```

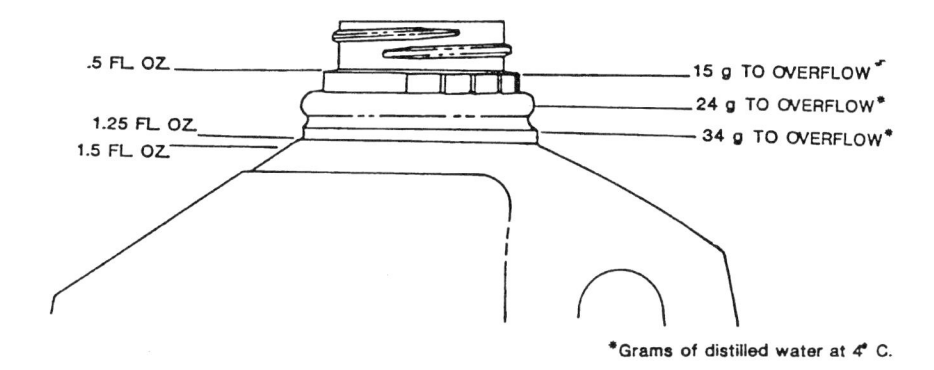

Figure 4.15 Lip cross-section, the finish shown is a standard 38 mm SP-400 finish with one turn of thread. Below the finish is a series of ratchets that engage a "tamper-evident" ring molded to the cap. Below the ratchet area is a bumper-roll that protects the "tamper-evident" ring from accidental damage. (Courtesy of Johnson Controls, Inc. Plastic Machinery Division)

flush volume control inserts; second, when containers are measured "fresh," a maximum of $\frac{1}{2}$ hour old, and with standard indented volume control inserts; and third, when containers are measured "fresh," and are intended to be used "fresh" without volume control inserts.

The second standard is 3805 g plus 60 g for two indented volume control inserts. The volume control inserts are designed to displace the difference in volume from shrinkage between "fresh" and "aged" containers (see Dow Chemical Co. Charts).

51.8 g	Weight of water displaced by two inserts
8.2 g	Weight added for safety factor
60.0 g	Weight standard for two inserts
3805.0 g	Overflow weight Standard I
3865.0 g	Overflow weight Standard II

The 3865 g standard is used when containers are measured "fresh" with flush volume control inserts or are measured without volume control inserts and are intended to be used "aged." This standard is the one most often used because most molds are fitted with volume control inserts and containers are usually measured "fresh" with flush inserts.

The standard for the half gallon dairy containers are as follows:

1891 g	Weight of $\frac{1}{2}$ gallon of distilled water at 4 °C (39.2 °F) to fill level 12.7 mm ($\frac{1}{2}$ in.) from top of container
15 g	Weight added for finished head space
9 g	Weight added for safety factor
1915 g	Overflow weight Standard I
24 g	Weight of water displaced by two volume control inserts
4 g	Weight added for safety factor
1943 g	Overflow weight Standard II

As with the gallon standards these half gallon standards are used in the same way under the same conditions.

4.4.3 Machine Line Mold Volume Correction

The difference in cubic volume between the measured containers and the standard is converted into height of the bottom split. Material is removed from the molds which will make the containers slightly shorter.

For example, suppose the measured volume of a container is the equivalent of 3965 g of water at 4 °C (39.2 °F). The containers have flush volume control inserts and are measured when half an hour old. The standard is 3865 g; therefore, 100 g must be removed. This is the equivalent of 100.07 cm^3 (6.109 in.^{3}). The area of the container at the bottom split is 221.68 cm^2 (34.36 in.^{2}). The following formula is used:

$$\frac{\text{Volume (grams)} \times \text{factor to in.}^3}{\text{Area at bottom split}} = \text{Height to be removed from bottom split} = 4.52 \text{ mm } (0.178 \text{ in.}) \text{ removed from height}$$

4.4.4 Source of Error in Volume Correction

Five errors, two intentional and three unintentional, accumulate to add a cushion to the volume corrections made on machine line molds.

- A tolerance of 9 g of water is added to the basic standard of the container.
- A tolerance of 8.2 g of water is added to the basic standard for gallon volume control inserts.
- The size of the volume control inserts approximates the amount of shrinkage that will take place between containers a few minutes old and containers several days old. When containers are measured "fresh," and "flush," as many are, some of that shrinkage factored in for the inserts already may have taken place. This inadvertently adds a small amount of volume to the containers.
- All measurements are made with water. Milk is heavier, but more important, the weight of milk above the filler valve will cause the container to bulge more. This inadvertency adds a small amount of volume to the containers.
- Containers, if not measured immediately, will "bulge" slightly from the weight of the fluid stretching the plastic material. This inadvertency also adds a small amount of volume to the containers.

These five errors add approximately 19.5 mm (0.75 fl oz.) to 0.03 l (1.00 fl oz.) of additional volume to gallon containers measured "fresh" with "flush" control inserts. Containers measured under other conditions or designed for other situations will have slightly less error. Drawing I shows the approximate fill level location of head space volume from 0.0148 to 0.0443 liters ($\frac{1}{2}$ to $1\frac{1}{2}$ fl oz.).

4.4.5 Package Dairy Mold Volume Correction

The sampling procedure is not used for dairy molds not a part of a filling machine line. Instead the actual cubic volume of the mold cavity is measured. As with container measurements, mold measurements have one of two standards for comparison. How the molds will be used and whether or not they have volume control inserts are the factors to be considered.

These standards, based on actual container measurements, approximately reflect typical container weight, shrinkage, and production conditions. Generally they are not as accurate as actual measurements.

4.4.6 Production Conditions

The volume error or cushion is left in the molds as a safety factor from changes in production conditions at the final plant location. It is virtually impossible to completely duplicate actual production conditions on the machine line at the moldmaker. Changes in conditions can easily add or subtract volume from the containers. A second volume correction is necessary after all production conditions have been stabilized and the dairy has become familiar with the equipment. Because of the cushion, the correction usually can be done quickly. The extra volume is removed from the mold bottom split line. If initially the molds were sized exactly, then this second adjustment could at times involve adding volume which requires a shim. Adding a shim to these molds is far more costly, Figure 4.16 illustrates a simple method of adjusting volume on an "in production mold." Also the dairy producing short containers is taking the risk of being caught by the "Weight and Measures" Department. This is one problem a dairy starting a new production line does not need.

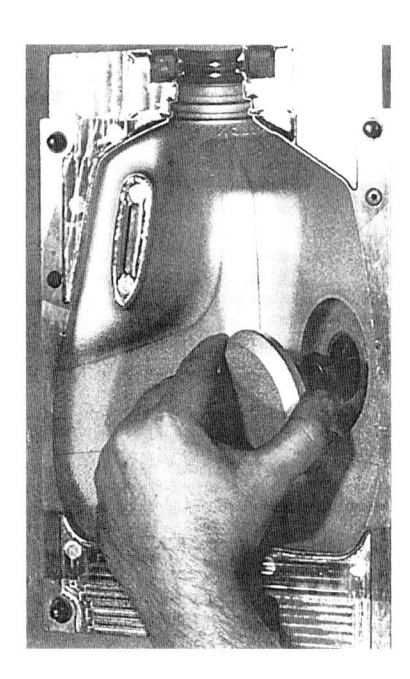

Figure 4.16 Interchangeable inserts can be used to adjust container volume. (Courtesy of Johnson Controls)

Six factors can change container volume (see Dow Chemical Operators Guide):

- Container weight,
- cycle time,
- blowing air pressure,
- melt temperature,
- mold temperature, and
- storage temperature.

The Dow Chemical charts illustrate these factors, plus the standard conditions under which the tests were made.

As the charts show changes in conditions, they can have a dramatic effect on container volume, for example. Suppose a dairy or molder intended to run a 70-g container at an 8-s cycle for storage in an unheated warehouse. Suppose further it is winter and the containers are held for approximately 10 days at a warehouse at approximately 5 °C (40 °F). The molds are volume corrected to these conditions. Later, because of increases in resin prices, the container weight is lowered to 60 g. Production requirements are slow, so cycle time is increased to 10 s. It is still winter and the containers are still held for 10 days. With these two changes the containers, by the charts, will be over volume by 0.037 liters ($1\frac{1}{4}$ fl oz). Production continues under these conditions into the summer. Warehouse temperature in the upper rafters is 60 °C (140 °F.) After 10 days the containers are (2 fl oz.) under volume.

This example is perhaps not very realistic; it does, however, show how changes in molding and storage conditions can change the volume of HDPE containers.

Two other molding conditions are suspected of altering container volume. Empirical data, however, are not available at this time to support the theory. First, exhaust time may be too short to allow the expanded blow air pressure to fully dissipate before the molds open. The extra pressure inside could stretch and distort the container which would add extra volume. The same reasoning can be applied to the leak detector. High pressures could also stretch and distort the container.

4.4.7 Conclusion

In dairy containers, achieving the exact volume target is a very complex process that must depend on several factors.

Maintaining control involves:

- Identifying all fundamental factors of time, temperature, and pressure that will affect volume
- Understanding the definition of volume, and not to become lost in a variety of conversion factors or so-called "equivalents" without knowing what they mean
- Holding production conditions consistent. The exact conditions really do not matter as long as they are consistent and precise.
- Measuring containers with the same style and technique. Again the exact method used does not really matter as long as it is consistent and precise.
- Correcting volume with adjustments to mold size, volume control inserts, or change in running conditions.

Figure 4.17 One Gallon Dairy Containers (Courtesy of Johnson Controls)

Figure 4.18 One Quart Dairy Containers (Courtesy of Johnson Controls)

Examples of bottles designed with the conditions discussed are shown in Figs. 4.17 and 4.18.

Acknowledgments and Notes

The bottle product design is based on the work of Chris Irwin, Johnson Controls.
    The section on container volume was developed by the engineering group of Johnson Controls and is a procedure used by the company.

The list in Section 4.4.6 is from Operators Guide—Controlling Shrinkage of HDPE Bottles, Dow Chemical, Midland, MI.

Note: Standards for neck threads for bottles can be obtained from Plastic Bottle Division, S.P.I., Washington, DC.

NEPCO is an acronym for Northern Engineering and Plastic Corporation. The company was one of the pioneer manufacturers of closures for the dairy industry and plastic milk bottles. Their original "screw-on" closure became an early standard that others copied, Blackhawk for example. Today there are a wide variety of screw-on and snap-on closure styles available. Courtesy of Johnson Controls, Inc. Plastic Machinery Division.

5
Industrial and Structural Part Design

In the last decade engineered blow molded parts have been developed and have had a significant impact in the automotive industry, large storage containers, transport packaging, and other industrial applications. Resins employed include not only traditional polyolefins, but so called engineered plastics such as ABS.

5.1 The Blow Molding Process

The blow molding process is ideal for large and thick parts because it has the capability of producing complex double wall parts in a single molding process. The knowledgeable designer or engineer will make proper use of pinch-offs, molded in stiffeners and structural ribs, providing a part with high molded stiffness-to-weight ratios and also giving a good moment of inertia for a given volume of resin.

The process affords much freedom in configuring outside opposing surfaces, which differs from one another. Cross-sectional structures are better than other processes. For instance the strength characteristics of a box and I beam forms can be duplicated. Inserts may be added for function or additional strength where required. Foam filling will also provide additional stiffness by creating a more homogeneous part.

5.1.1 Preferred Process

High production plastic parts that are hollow such as in tanks, pressure vessels, air ducts, cable channels, automotive components, and municipal carts, blow molding is often the preferred process, particularly when the quantities justify machine and mold expenditures (Fig. 5.1).

5.1.2 Hollow Parts

The hollow construction of blow molded parts provides the highest moment of inertia for a given volume of resin, compared with other processes such as injection molding and reaction injection molding, which are single shelled.

60 Industrial and Structural Part Design

Figure 5.1 Array of industrial blow molded products (Courtesy of Krupp Kautex)

5.1.3 Resin/Fiberglass Layup and Structural Foam Molding

For large structural parts resin fiberglass layup has been used along with other processes such as reaction injection molding (RIM) and various types of compression molding. With the blow molding process double-walled parts are made in a single cycle, making it cost effective.

With the proper use of tack-offs, structural ribs, foam filling, or molded-in stiffeners, it provides the highest as molded stiffness-to-weight ratios. This will be illustrated when a tack-off rib section is compared with the structural properties of an "I" beam (Fig. 5.2).

5.1.4 Foam-Filled

In addition the inside of a hollow blow molded part can serve as the outside skin of a composite structure. The inside, when filled with a polyurethane foam, or even expanded beads of polystyrene foam, forms the core of the composite and adds structural rigidity, insulation, and/or buoyancy to the finished part (Fig. 5.3). The availability of engineering resins has now allowed blow molded hollow parts to meet demanding requirements of products exposed to high heat, internal pressures, heavy loading, and severe impact. Thus parts can now be efficiently and economically produced that can store, transport, control, and distribute fluids, gas, powders, and other fluids as well as automotive components (see the plastic spoiler in Figs. 5.4a and 5.4b).

5.1 The Blow Molding Process   61

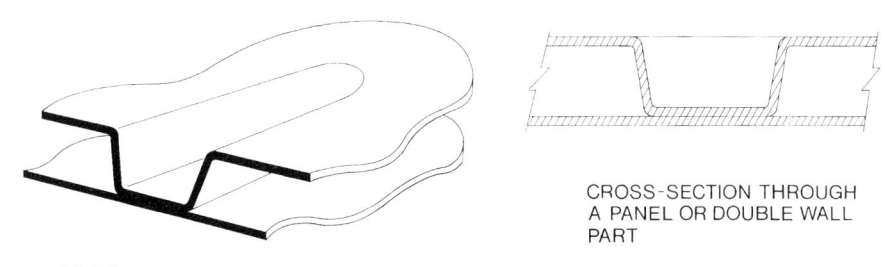

Figure 5.2 I beam

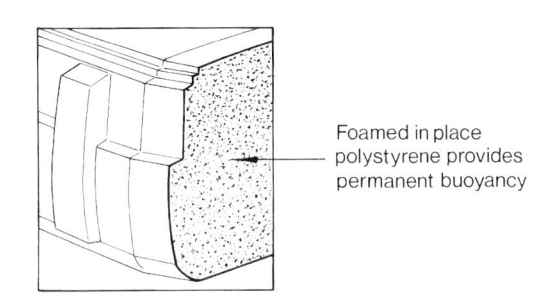

CROSS-SECTION THROUGH
A PANEL OR DOUBLE WALL
PART

Foamed in place polystyrene provides permanent buoyancy

Figure 5.3 Foamed dock float (Polyfloat) (Courtesy of White Ridge, Inc.)

Figure 5.4(a) Molding of spoiler (Courtesy of Plastic Omnium)

Figure 5.4(b) Blow molded spoiler (Courtesy of Plastic Omnium)

5.2 Kinetic Energy Design Engineering

The blow molded part is ideal for producing parts that have to absorb and dissipate mechanical energy, generally from impact.

The double-walled construction lends itself to several systems of kinetic energy dissipation.

5.2.1 Energy Management Concepts

5.2.1.1 Deformation (of the Surface Skin in the Area of Impact)

Because of the hollow nature of the part it deforms into itself at the point of impact and recovers its original shape.

5.2.1.2 A Sealed Hollow Part

A sealed hollow part is prepressured which distributes the absorbed energy due to decrease of volume. By properly sizing the orifice (a little hole) the part will be vented and will control the amount of damping (Fig. 5.5).

5.2.1.3 Foam-Filled

A blow molded hollow part may be designed to dissipate considerable mechanical energy through planned crushing of the structure or filling between the opposite side skins made

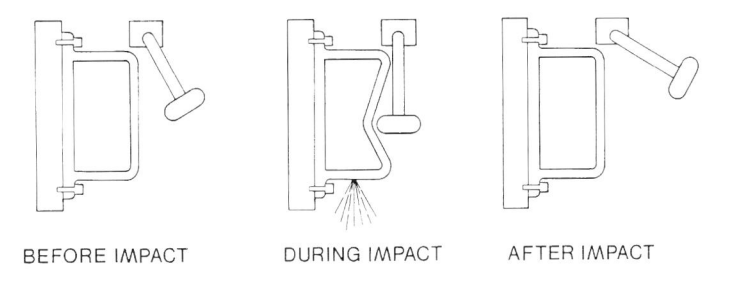

Figure 5.5 Engineered blow molded bumper

from engineered blow molded materials. This technique is used for bumpers in the automotive and transport industries (Fig. 5.6).

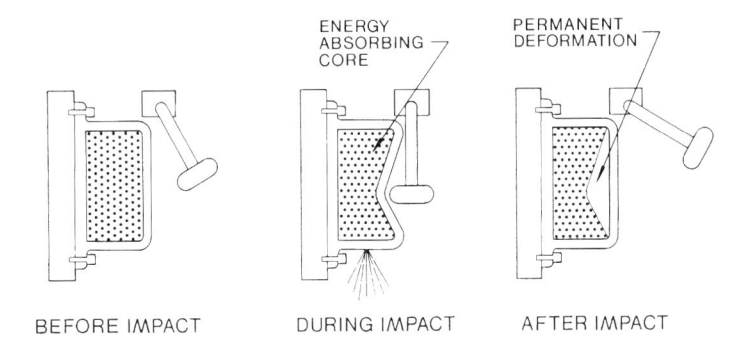

Figure 5.6 Foam filled EBM bumper

5.2.1.4 Crushing

When the interior is filled with a foam material, if it is rigid, it will cause nonrecoverable deformation, but if filled with flexible material then it will absorb the impact.

5.3 Molded-In Insert of Components

Blow molded parts have few restrictions regarding the shape, size, material, and support during the molding process because of the low pressure required during the blowing of the parison (Fig. 5.7). (Also see Fig. 10.19.) When irregular shapes are used the blow parison forms a tight fitting wall of plastic around the insert. *Note of caution:* Post assembly of screw inserts is often preferred by manufacturing groups; however careful the production operation, inserts always find their way into "regrind" and consequently fine pieces of

64   Industrial and Structural Part Design

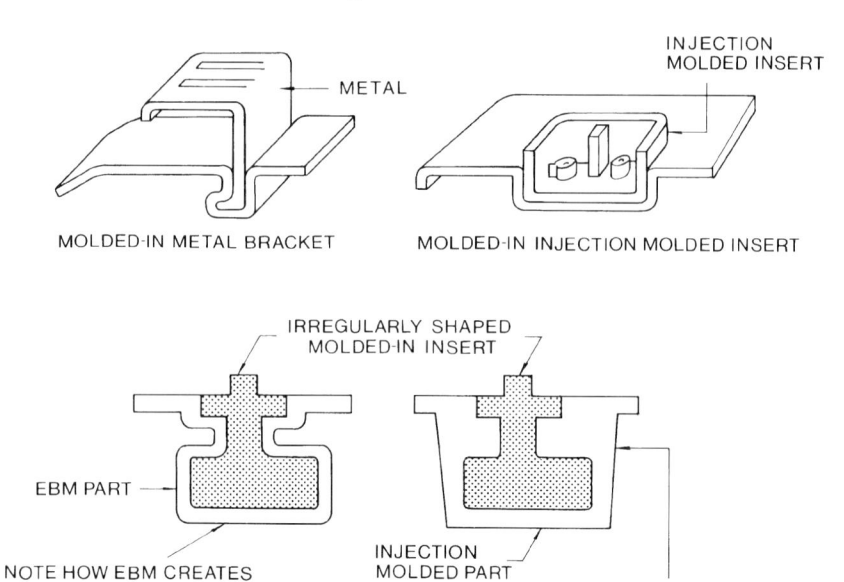

Figure 5.7 Molded-in details

insert pass through the extruder and into the parison. It does not require much imagination to see the havoc this can cause in production. An example of the ultimate use of inserts is in the automotive gas tank (Fig. 5.8). Figure 5.7 shows molded-in details.

Figure 5.8 Gasoline tank with internal barrier inserts (not shown) molded in tank during process (Courtesy of Plastic Omnium)

5.4 Interlocking Systems

This technique is based upon the interference fit between the male and female configurations of the interlock. The locking detail or detent prevents accidental disassembly. It is used to create snap together assemblies and panel systems (Fig. 5.9). In the mold construction where this type of assembly is desired the cavity is cut "metal safe" so that the parts fit together with clearance. Metal is then removed for the mold to tighten up, the interlock/snap fit, after sampling, repeating several times until the desired fit is obtained. Variations of the basic design features of the interlock may be explored.

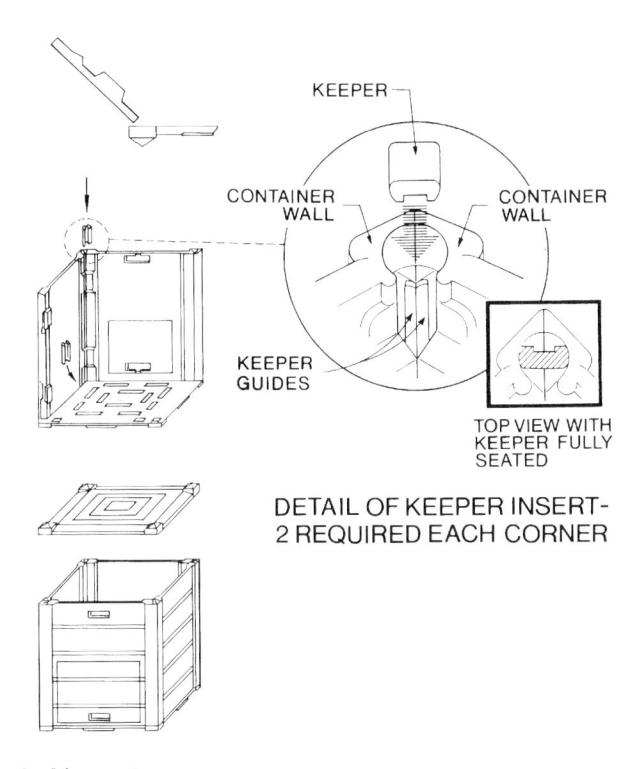

Figure 5.9 Interlocking systems

5.5 Snap Fits

Open top containers with lids (such as trash cans) may be molded with the container and lid in one piece and cut apart in a post molding operation (Fig. 5.10). When the parts are cut apart as in trash cans the lid will "snap-fit" to the container lip as seen in Fig. 5.11.

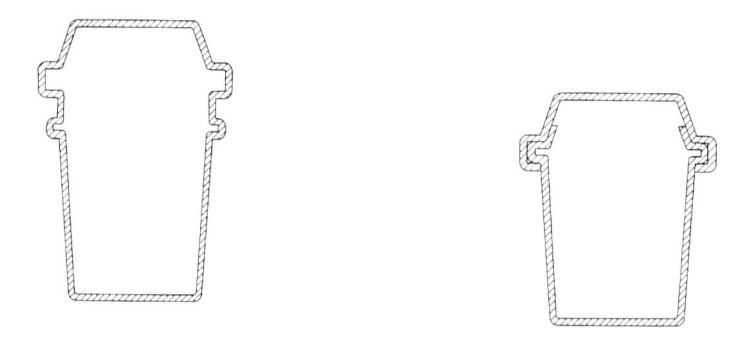

Figure 5.10 Two-part-one piece

Figure 5.11 Snap-fit

5.6 Multiple/Combination Cavities

To improve the economics of producing an open top product, making two parts from the same parison is commonly considered. In "two-up" molding, the mold is built with the open ends together, and the molded parts are cut apart. A short transition between the parts is desirable so the part may be blown through a needle which pierces and blows through this area. No hole is left in the part, since two cuts are made separating the containers and transition (flash), later being "reground" in the same technique used with a container and lid (Fig. 5.12).

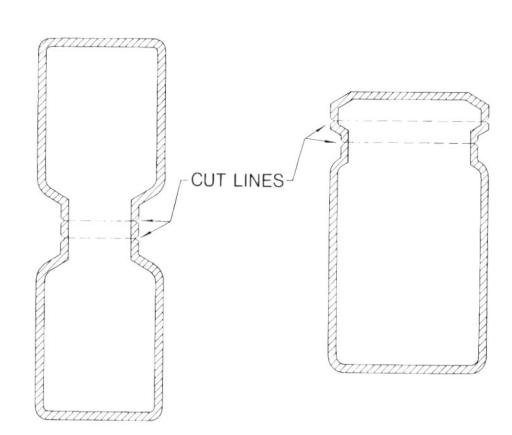

Figure 5.12 Two-up molding

5.7 Container Configuration Design

The effects of warpage on open top containers that are square or rectangular shape are considered in this section.

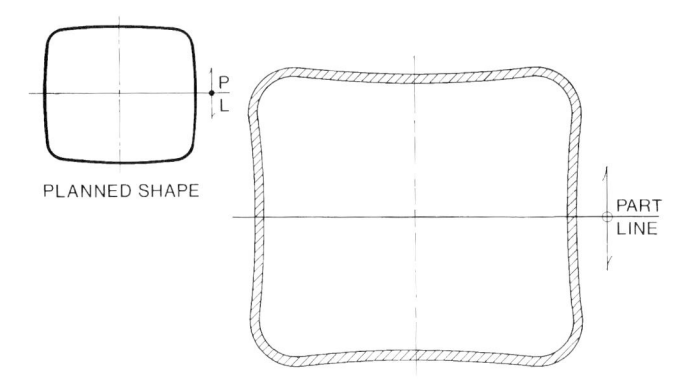

Figure 5.13 Effect of warpage

5.7.1 Flat Sides

Flat sides tend to warp as shown in Fig. 5.13. This warpage is more exaggerated after the part is cut owing to stress setup by shrinkage variations. A shallow dome shape which tends to control the shrinkage direction is shown in Fig. 5.14.

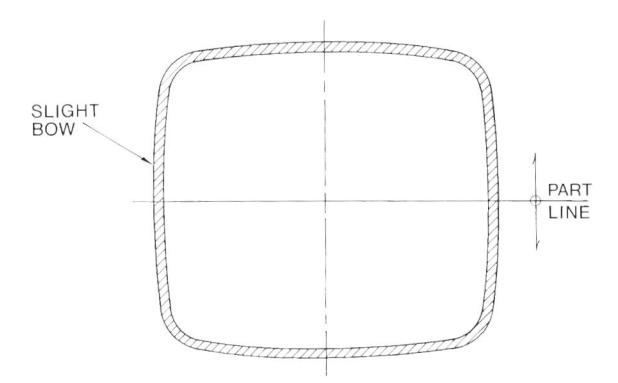

Figure 5.14 Ideal shape

5.7.2 Lip

The opening of the container is usually where the lip configuration occurs. This is an area where warpage usually takes place. The avoid warpage lips should also have a cross-section to counter the warpage (Fig. 5.15). Various configurations are shown. Care should be taken not to violate the blow ratio, that is, depth is one half the width.

68 Industrial and Structural Part Design

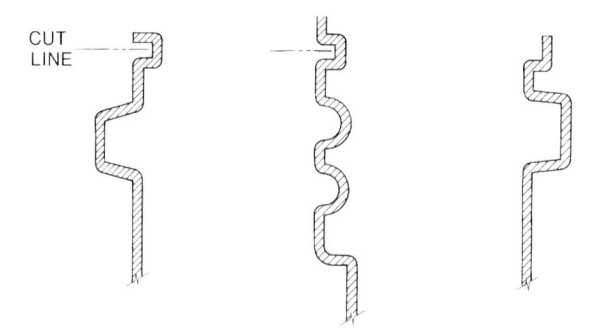

Figure 5.15 Lip configurations

5.7.3 Nesting and Stacking

For economy of storage and shipment the container part should nest to provide maximum stacking (Fig. 5.16a to c).

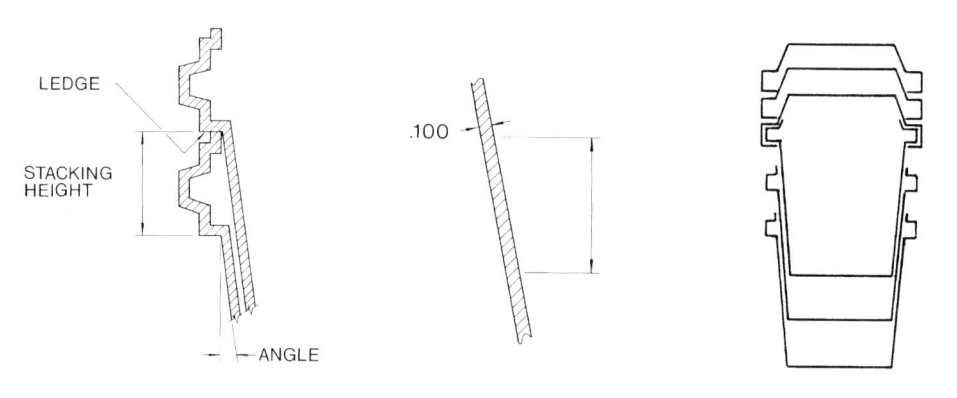

Figure 5.16(a) Nest and stacking

Figure 5.16(b) Taper

Figure 5.16(c) Stacking height

The lip must clear ledge of upper part. The nesting is a function of stacking height and angle of side wall of taper. The angle of side wall is dependent on wall thickness of side of container.

$$1° = 0.0175'/\text{in. wall thickness.}$$

(Stacking angle at 1 in. stacking height)

Stacking height is a compromise with nesting angle (Fig. 5.17).

5.7 Container Configuration Design   69

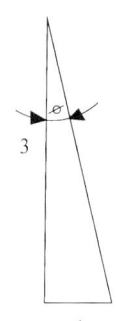

Figure 5.17 Nesting angle

Example:

$$3 \text{ in. stacking height} - 0.100 \text{ wall}$$

$$\tan \phi \frac{0.1}{3} = 0.0333$$

from trigonometric tables = 1°55′.

Therefore the stacking angle is 2°. For example see Fig. 5.18.

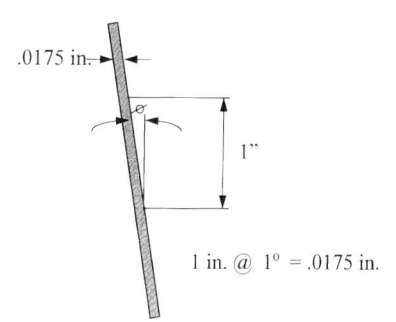

Figure 5.18 Stacking angle

5.7.4 Cutting Containers Apart

A saw cut may leave a fuzzy edge which needs to be trimmed, usually with a knife or gas flame.

It is most desirable to cut with a knife blade in a fixture. A groove to guide the cutter blade is desirable, otherwise the top lip will be wavy and uneven (Fig. 5.19).

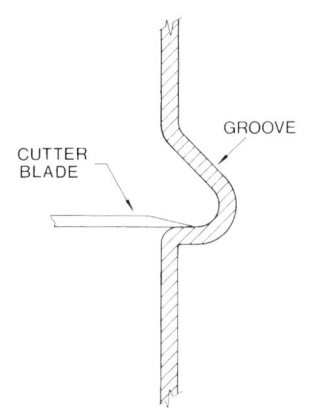

Figure 5.19 Cutter groove

5.8 Conclusion

When designing new plastic parts, it is important to take full advantage of design and process opportunities to increase the functional value of the part. Structural and functional features, as well as environmental enhancements, may be added through design and material selection.

Most automotive instrumental retainers, such as fixtures to attach the defroster, and heating, ventilating, and air conditioning components, among other items are injection molded (Fig. 5.20). But with blow molding, a structure can be designed to incorporate the

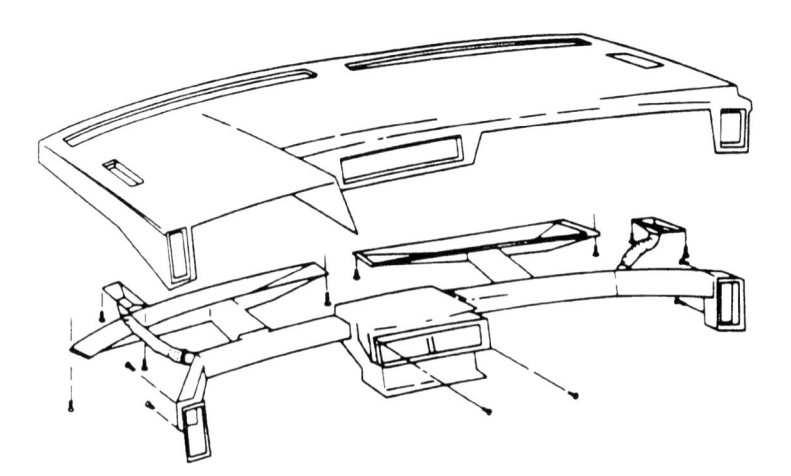

Figure 5.20 Injection retainer with air distribution system (Courtesy General Electric)

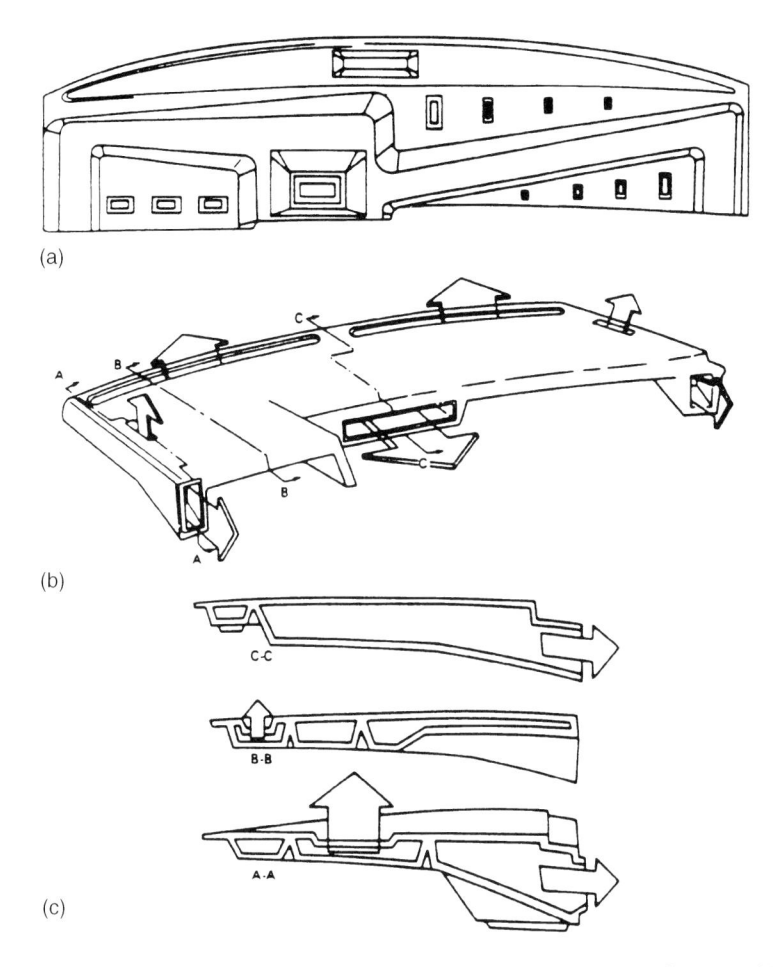

Figure 5.21 Blow molded instrument panel upper retainer (a) top, (b) bottom, (c) cut-away (Courtesy of G.E. Plastics)

air distribution system in the hollow walls of the part (Fig. 5.21). The design of the blow molded part reduces the number of parts and the number of assembly steps, adds stiffness by means of structural tack-offs, and increases the space available inside the instrument panel [1].

Acknowledgments

The material in this chapter is drawn from the work of the GE Plastic Study of Engineered Blow Molded Plastic.
   Author of "Engineered Blow Molding Part Design", Lincoln J. Alvord.
   Note: Suggested Voluntary Thread Specification for Plastic Drums. Contact Plastic Drum Institute S.P.I., Washington, DC

5.9 Reference

1. Ferfunson, L. and Taylor, B., *Blow Molding*. In Dostal, C.A. (Ed.) *Engineering Materials Handbook, Vol. 2, Engineering Plastics*, A.S.M. International, Metals Park, OH

6

Computer Aided Design and Engineering Analysis

The need to accurately predict part wall thickness, blow ratio, and thinning is apparent from the previous discussion of "Rules of Thumb" and guidelines for the engineering layout of the basic geometry. In addition the following guidelines usually result in a moldable part without excessive blowouts or other factors affecting polymer thinning.

6.1 Performance Criteria

For industrial parts with performance criteria, or bottles and containers where material is an important cost factor, the ability to quantitatively predict the wall thickness distribution either locally or throughout the part is essential. There is a need to understand both quantitatively and qualitatively how and where the material thins. Also important are geometric factors such as shape, blow ratio, draft angle, and radii, all of which affect the final part.

Finite element analysis software has been developed to address these issues.

6.2 Computer Software Simulation

Computer software simulation of a molded parison is in its infancy. Several commercial programs are available that are based on process models, liquid flow, or solid deformation (See Fig. 6.1) [1].

The extrusion blow molding parison can be simulated by computer software programs that use liquid flow theory, and predict the shape and thickness of the extruded polymer prior to pinch-off. Then, once the polymer is "pinched-off," its blowing behavior is simulated. This simulation provides a valuable understanding of the part, parison, and mold design prior to tool manufacturing. It provides data as to whether the part will achieve wall thickness objectives, structural requirements, and optimized molding cycle. Further, an important feature is the ability to have simulated data on the wall, blow ratios, and cooling

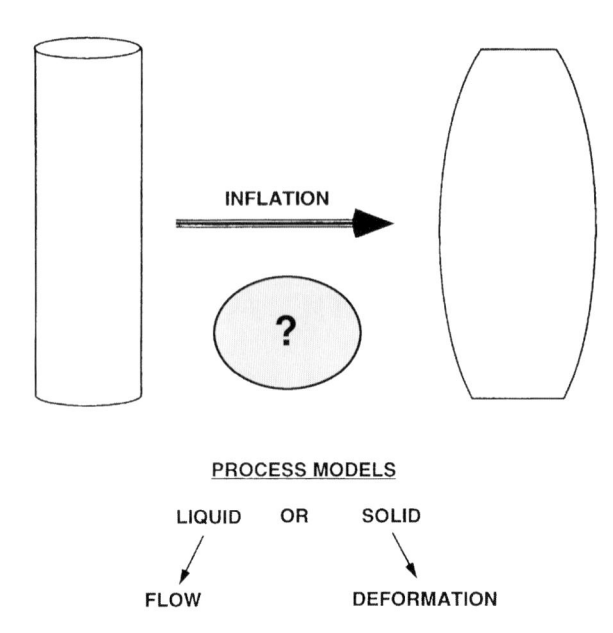

Figure 6.1 Overview of blow molding (Courtesy of LR Schmidt Associates)

requirements, which provide guidance for parison programming or heater placement to achieve optimum product quality [2].

In summary, the entire simulation process may also predict the following conditions:

The polymer temperature resulting from cooling of the die
The effects of gravity or sagging of the extruded parison
The final wall thickness and flow rate (melt index) of the extruded polymer
The process sensitivity to operating parameters.

6.3 Reducing Parison Thickness

C-mold software, which uses the liquid flow theory, is a commercially available product. The objective is to illustrate the project based on an existing extrusion blow molded bottle for dish washing liquid, and to reduce the original extruded parison thickness from 0.228 mm (0.090 in.) to 0.152 mm (0.060 in.). The challenge was to determine if the final thickness would still meet product specifications and to predict how much cooling time would be reduced.

In Fig. 6.2a the C-MOLD simulation shows the predicted final wall thickness for both starting parison thicknesses. The thinnest areas are in the corners of the bottles—approximately 0.048 mm (0.019 in.) for the original parison and 0.038 mm (0.015 in.) for

6.3 Reducing Parison Thickness   75

| 0.090
0.070
0.050
0.030
| 0.010

Original Parison
Thickness 0.090"

| 0.090
0.070
0.050
0.030
| 0.010

Original Parison
Thickness 0.060"

| 17.0
12.8
8.7
4.6
| 0.5

Original Parison
Thickness 0.090"

| 17.0
12.8
8.7
4.6
| 0.5

Original Parison
Thickness 0.060"

| 5.0
4.0
3.0
2.0
| 1.0

Original Parison
Thickness 0.090"

| 5.0
4.0
3.0
2.0
| 1.0

Original Parison
Thickness 0.060"

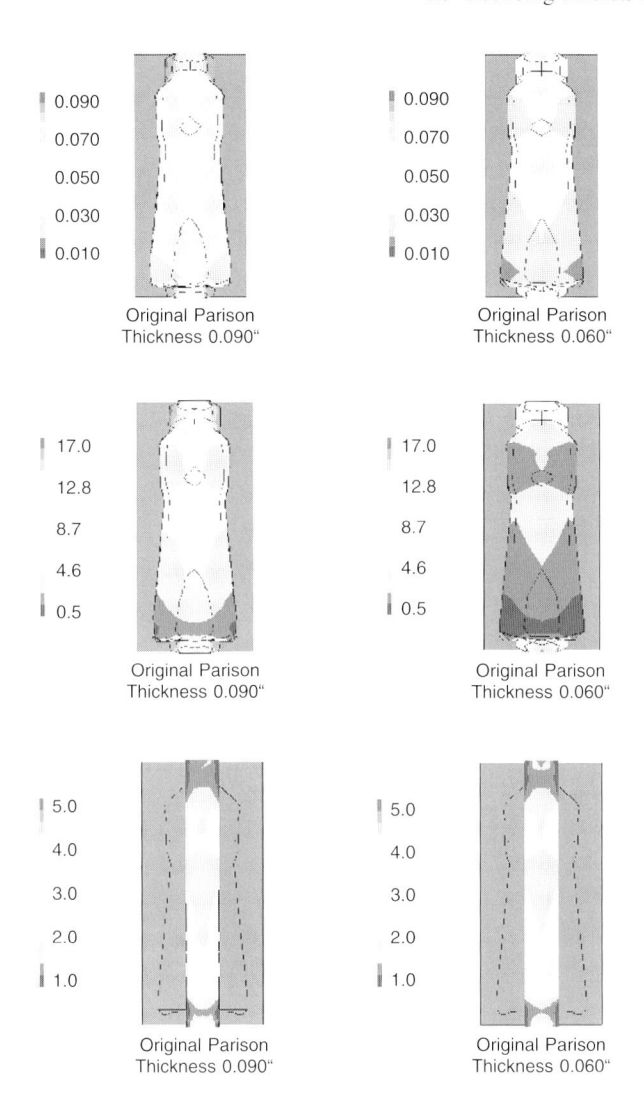

Figure 6.2(a, b, c) Blow molding: wall thickness distribution (Courtesy of A.C. Technology, Louisville, KY)

the reduced parison—a wall thickness reduction of about 21%. Because the thick area near the cap does not stretch significantly, it dictates the cooling time. With the 0.152 mm (0.060 in.) parison, faster cooling at the pinch-off area should allow the bottle to be ejected sooner. Because cooling time is proportional to the square of thickness, the 0.152 mm (0.060 in.) parison should give a better cycle time with a relatively small trade-off in the wall thickness at the corners.

Figure 6.2b shows that the 0.152 mm (0.060 in.) parison has a considerably lower cycle time. The estimated cooling time can be reduced from 17 s for the original parison to 7 s for the new design—a 59% reduction that has a significant impact on production rate.

The software can also display the distribution of area/stretch ratio, that is, how much an area of the parison stretches relative to its original surface. Areas at the corners stretch significantly more than the other areas of the part. This stretch data can be mapped back to the original parison (Fig. 6.2c), indicating very precisely how to modify the profile of the 0.152 mm (0.060 in.) parison to achieve optimally uniform wall thickness. Stretch data can be mapped back to the area/stretch ratio to allow the parison to be programmed to be thicker in selected areas. For example, if the part specification has required that the corners remain thicker, the parison could be programmed thicker in the area that blows into the corners without reverting to a thicker parison for the whole part. Although it was not included in the analysis in this case, modifying the thinned down parison profile with the aid of mapped-back area/stretch data could have provided the optimal balance between material savings and part quality.

6.4 Fluid Flow Finite Element Simulation

Another software program, "Polyflow," is based on a fluid flow where the material is modeled as a finite element simulation using a generalized Newtonian fluid theory [3].

6.4.1 Modeling

The fluid is drawn in Fig. 6.3, where a cross-section through the mold and the fluid parison is shown. In this process, the fluid parison will inflate under the influence of a blowing

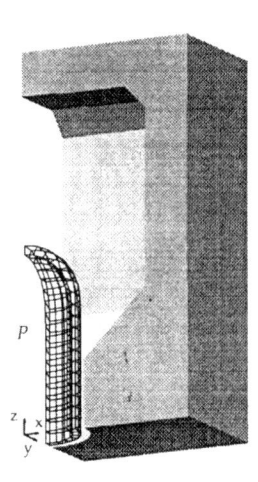

Figure 6.3 Three-dimensional mold and finite element mesh in the parison (Courtesy of Fluent, Inc. Lebanon, NH)

pressure. During inflation, the parison thickness progressively decreases until the fluid hits the wall of the mold. At the contact, the velocity component vanishes, while slipping can still be considered in a tangential direction.

6.4.2 Simulation

The following is the numerical simulation of the blow molding process sketched in Fig. 6.3 for a typical three dimensional case. The two views of the mold geometry are displayed in Fig. 6.4. It is noted that the geometry is characterized by two planes of symmetry, which are taken into account in the calculation. The inner radius of the neck of the vessel is 1.5 cm.

The initial finite element mesh for the fluid parison is given in Fig. 6.3. The number of elements and nodes in the fluid region will actually remain unchanged throughout the calculation, as the remeshing algorithm affects only the geometry of the elements. The initial parison is characterized by a uniform thickness distribution. From a geometric view point, the inner radius of the initial parison is 1 cm. At this level also, the initial shape of the parison is not yet optimized. However, tracking of material particle trajectories would give useful information for optimizing the initial parison shape.

The fluid velocity equals 2104 Pa/s for the blowing pressure $P = 105$ Pa. In Fig. 6.5, the numerical results of the inflation process are displayed at several intervals. At first, it is found that the blowing time equals about 0.15 s. In the present case, it is observed that the inflation occurs in the horizontal direction first and then the vertical direction. The inflation is very slow at the beginning of the process and accelerates at the end as the result of the combined effect of the constant pressure and of the thickness reduction.

Now observe the contact between the fluid parison and the mold surface. The first contact is made at the neck of the vessel. Here, the thickness of the blown product is

Figure 6.4 Geometry of the three-dimensional mold (Courtesy of Fluent, Inc. Lebanon, NH)

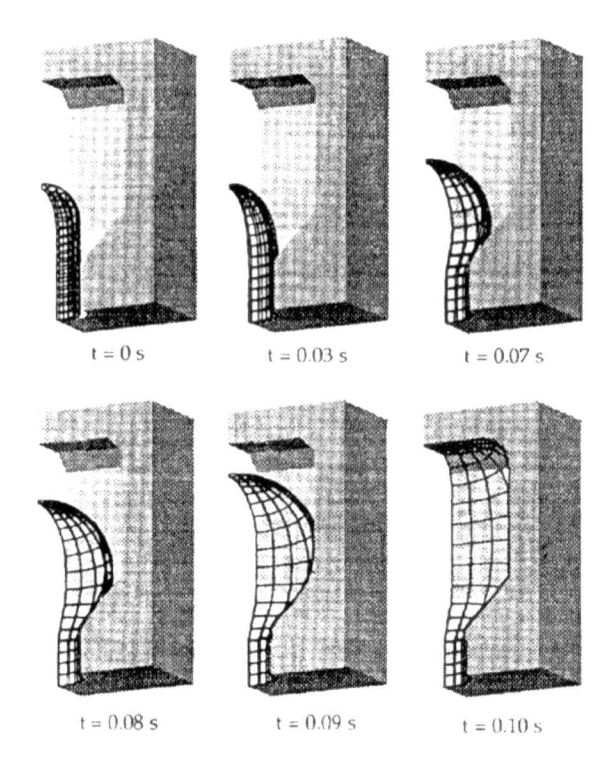

Figure 6.5 Blow molding simulation of a Newtonian fluid (Courtesy of Fluent, Inc. Lebanon, NH)

relatively important, and will remain essentially unchanged, in view of the no slip boundary conditions of the fluid at the surface of the mold. When the inflation is in progress, the motion is accompanied by thickness reduction. At the end of the process the thickness at the bottom of the part is low. This is a direct result of the no slip boundary condition of the fluid when the contact occurs.

In Fig. 6.6 the parison shape at time intervals in two cross-sectional planes of symmetry is displayed. The parison's motion, together with the thickness, changes as a function of time.

This development of an efficient mechanical technique combined with a robust remeshing algorithm has allowed the three-dimensional numerical simulation of the blow molding process. The blowing time as well as the development of flow kinematics and the final thickness is predicted and the distribution of the blown product is calculated, and given an initial shape.

6.4.3 Prediction Example

An example of the application of this Polyflow software is shown in Fig. 6.7. The program predicts the shape and thickness of extruded polymer prior to pinch-off. Once pinched-off,

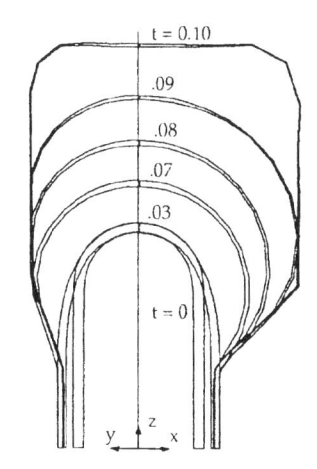

Figure 6.6 Blow molding simulation. Parison shape in the planes of symmetry, at various times (Courtesy of Fluent, Inc. Lebanon, NH)

it simulates the blowing process of the polymer. Furthermore, for the entire process it is able to predict the following:

the polymer temperature resulting from die cooling,
the effects of gravity sagging on the extruded parison,
the final wall thickness of the product given the thickness and flow rate of the extruder polymer, and
the process sensitivity to operating parameters.

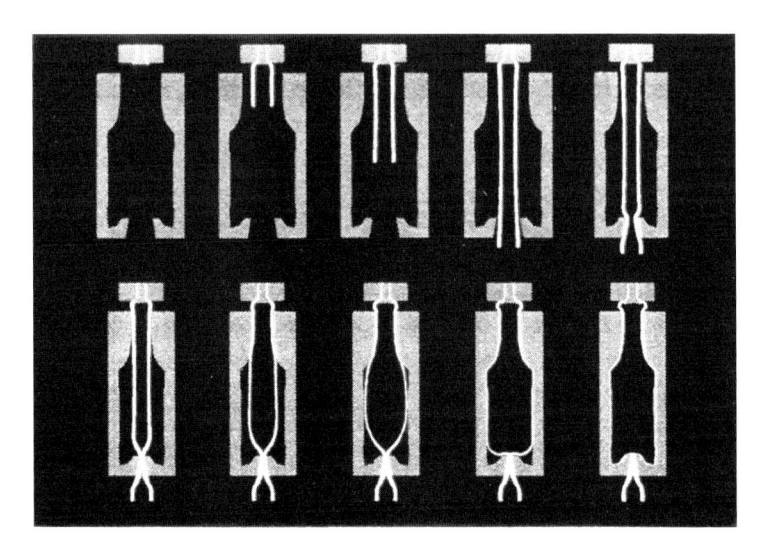

Figure 6.7 Polyflow prediction of shape and thickness (Courtesy of Fluent, Inc. Lebanon, NH)

6.5 Polymer Inflation and Thinning Analysis [4]

An additional software program, PITA (Polymer Inflation and Thinning Analysis), has been developed using solid deformation theory. This software, which uses the dynamics of parison inflation and mold closing, has been proven to be a reliable predictor of wall thickness distribution in actual blow molded parts. Typically, finite element meshes are created for the parison and the cavity surface of each of the mold halves. Also, relevant process data, such as internal parison pressure versus time and extrusion velocities, are input into the model. The resultant simulation will give the wall thickness values at any location of the blow molded parts. These are then plotted graphically and/or input to a structural analysis software or program.

6.5.1 Geometric

By examining the effects on common occurring geometries, known as primitive shapes, wall thickness distribution may be understood. In some cases a detail from an actual part under consideration can be approximated by one of these shapes and the designer/engineer may examine the curves to estimate wall thickness distributions.

6.5.2 Understanding Wall Thickness

To understand the meaning and usage of these wall thickness distribution curves, consider the blow molding of a simple cylindrical projection as shown in Fig. 6.8, then in Fig. 6.9. A cross-section of the right circular cylindrical (female) cavity is necessary to produce this projection. The angle will be used to define the particular position under consideration on

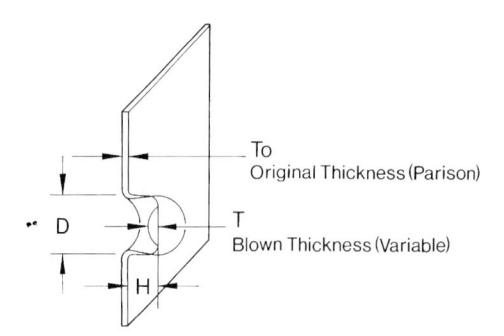

Figure 6.8 Blown cylindrical projection (Courtesy of A.C. Technologies, Louisville, KY)

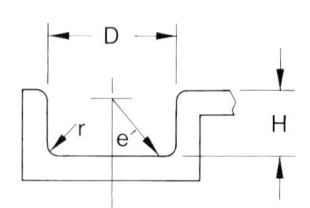

Figure 6.9 Mold cavity to form right cylindrical projection (Courtesy of A.C. Technologies, Louisville, KY)

6.5 Polymer Inflation and Thinning Analysis [4]

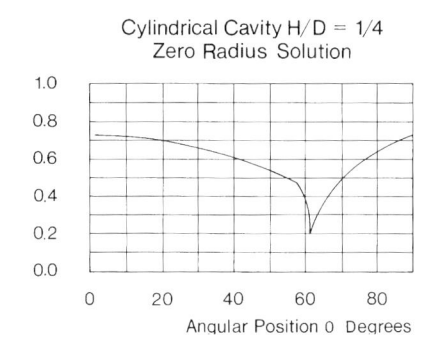

Figure 6.10 Zero radius solution wall thickness distribution curve for $H/D = 1/4$ (Courtesy of A.C. Technologies, Louisville, KY)

the surface of the cavity, and the same position on the blown part. For example, at the top edge of the cavity $= 90°$, while at the bottom center of the cavity $= 0°$. Now consider the special case where $H/D = 1/4$ (or $BR = 0.25$), and there is no corner radius ($r = 0$). The PITA modeling can develop a curve showing the actual wall thickness distribution based on angle as shown in Fig. 6.10. Note that thickness is presented as the "normalized thickness" and is the ratio of the blown wall thickness (T) to the original parison thickness (T_0). This particular curve is referred to as the "zero radius solution." When radius is added to the bottom corner of the cavity, a more realistic situation is created. Note that for the purpose of curve generation and presentation, the corner radius is always considered in terms of its ratio to the cavity diameter (r/D). Additional curves for $r/D = 0.10$ and $r/D = 0.20$ have been added to the zero radius solution ($r/D = 0$), and are shown in Fig. 6.11. It is interesting and significant to note that the new curves coincide with the original $r/D = 0$ curve in all areas outside of the radius area. In addition, the wall section within the radius can be approximated as nearly constant. This is significant since it greatly simplifies the use of this powerful tool as only the zero radius solution needs to be given. Consider the following examples.

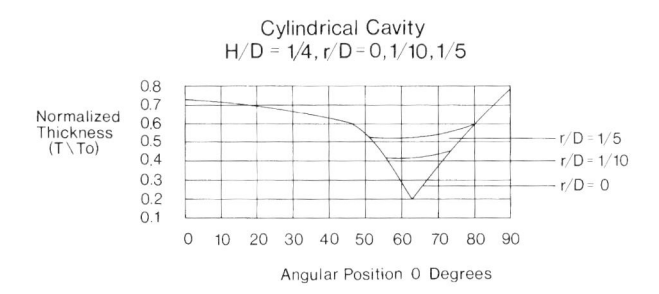

Figure 6.11 Wall thickness distribution curve for various H/D ratios (Courtesy of A.C. Technologies, Louisville, KY)

6.5.2.1 Using Normalized Thickness Curves

The curve for a right circular cylinder with $H/D = 1/4$ will again be used to demonstrate the application of the PITA graphs. For the purpose of these examples, the following dimension will be assumed:

Diameter	$D = 25.4$ mm (1.0 in.)
Height	$H = 6.35$ mm (0.25 in.)
Corner radius	$r = 0.787$ mm (0.031 in.)
Parison thickness	$T_0 = 3.175$ mm (0.125 in.)

Example 1: Find the minimum wall section in the projection in Fig. 6.12. Calculate using the relationship: $= \tan 1[1/H(D/2 - r)] = 62$.

Locate the value of the horizontal axis of Fig. 6.10 and project up to the curve, and read T/T_0 on the vertical axis:

$$\text{For } = 62, \ T/T_0 - 0.28$$

Now solve for the actual thickness:

$$T = 0.28(0.125) = 0.035 \text{ in.}$$

Since the wall section is nearly constant along the arc of a radius, this is considered the thickness within the edge radius. Since a 0.889 mm (0.035 in.) thickness is relatively thin, the design engineer should consider increasing the radius.

Example 2: (Fig. 6.13) If a 9.525 mm (3/8 in.) diameter hole is punched in the center of the top surface of the projection, what is the wall thickness at the cut?

Find for $b = 3/16$ in. from relationship III of Fig. 6.14:

$$= \tan 1 b/H = 37$$

Locate $T = 37$ on the horizontal axis of Fig. 6.10 and project up to the curve and read normalized thickness on the vertical axis:

$$\text{For } Q - 37, \ T/T_0 = 0.65$$

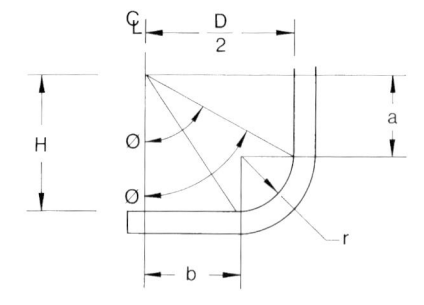

Detail & Relationships for Mold Cavity to Create Right Cylindrical Projection

Figure 6.12 Minimum wall section in the projection

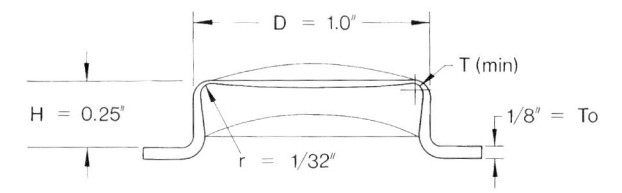

Figure 6.13 Minimum wall section with hole

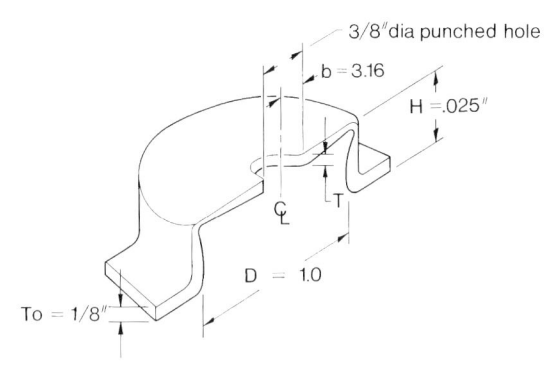

Figure 6.14 Detail and relationship for mold cavity to create right cylindrical projection

Solve for thickness at cut, T:

$$T = 0.65(T_0) = 0.65(0.125 \text{ in.}) = 0.081 \text{ in.}$$

6.6 Conclusion

Computer aided design and engineering analysis has become very important in designing moldable parts. By using such programs one can accurately predict the part's wall thickness, blow ratio, and thinning and produce parts with fewer blowouts. Computer Simulation, Fluid Flow Finite Element Simulation, and Polymer Inflation and Thinning Analysis are all such programs that a designer can use to help in producing a moldable product.

6.7 References

1. Schmidt, L.R. *Blow Molding Analysis and Process Simulation1*, L.R. Schmidt Associates, 11 Front Street, Schenectady, NY 12305-1312
2. Spann, J., *C-MOLD*, A.C. Technology, Louisville, KY
3. Polyflow, Fluent Inc., 10 Cavendish Court, Lebanon, NH 03766

4. Spann, J., PITA-A.C. Technology, Louisville, KY
5. Debault, D., HocQ, B., Jiang, Y., *3-D Numerical Simulation of the Blow Molding Process* and Marchal Polyflow S.A., Place de l'Université 16, B-1348 Louvain-La-Neuve, Belgium
6. Alvord, L.J., *Engineered Blow Mold Part Design*, G.E. Plastics. Pittsfield, PA

Notes

C-PITA is based on blow molding and thermoforming simulations developed by GE Corporate Research & Development and extensively verified by GE Plastics. AC Technology holds an exclusive license to commercialize and market this software. C-PITA development is supported by a consortium of industrial companies, industry experts, and educational institutions. This collaborative effort, with feedback and direction from potential end users of the technology, has made C-PITA a reliable, user friendly analysis tool for the blow-molding and thermoforming industries. C-MOLD is a set of integrated CAE simulation for plastic molding process. Analysis modules address injection molding, including coinjection molding, gas-assisted injection molding, and part shrinkage and warpage analysis; reactive molding processes; and blow molding and thermoforming. C-MOLD analyses provide answers at all stages of the design and manufacturing process to identify workable solutions, improve productivity, and enhance part quality. C-MOLD is used in all areas of the plastics industry. Significant market segments include polymer suppliers, auto makers, electronics manufacturers, medical device manufacturers, mold and tool makers, and molding machine manufacturers.

7
Decoration of Blow Molded Products

7.1 Introduction

Today products are decorated for customer appeal, which helps people decide what they would like to buy. In many instances the decoration indicates what is in the container or how to use the product. There are several ways to decorate blow molded products, such as:

- Labels—both post molding and in mold,
- decals,
- screen printing,
- hot stamping,
- pad printing, and
- paint.

Each process is specialized, has a number of options, and many do's and don'ts. There are too many to cover completely within the scope of this chapter, particularly with generalizations and rules of thumb. The decoration of plastic surfaces has been covered in detail in books specifically devoted to the subject [1–3] and discussed in books covering other aspects of plastic technology [4–6].

7.2 Surface Treatment

To be able to coat many plastics, the coating, whether glue, ink, paint, etc., must adhere to provide a useful and functional part. Thus a requisite step in most decoration of plastic, especially polyolefins, is surface treatment. This involves changing surface tension of areas to which decoration is to be applied. A useful way to determine whether treatment is needed is a water test—when water is spread over the plastic surface, a surface that is well treated will wet out. With a smooth layer of water on a poorly treated or untreated surface, the water will bead. An example is how water beads on freshly waxed cars, which acts much like an untreated surface. Another simple test that can be performed is with commercially available marking pens, whose inks are available in differing ranges of surfaces tension.

7.2.1 Surface Treatment Methods

There are several methods to treat plastic surfaces. The most common one for treating polyolefins is flame treatment in which the surface is exposed to a gas flame. An alternative treatment is corona, in which the part is subjected to an electric discharge. More commonly used for containers, this treatment is not often used in large part molding. Beyond those for polyolefins, the other techniques that are used to surface treat include:

- Washing with water-based chemicals,
- solvent cleaning and etching,
- mechanical abrasion—sanding,
- chemical etching, and
- additives compounded into resin

7.2.2 Flame Treatment

With flame treatment the surface is oxidized and it becomes easily wettable. The oxidizing portion of the flame is in contact with the plastic surface for a period of less than a second. Flame treatment is frequently the method of choice for treating irregularly shaped objects. See Figs. 7.1, 7.2, 7.3, and 7.4 for adaptability of such products [5].

7.2.3 Corona Discharge

This process consists of passing the polymer surface through a plasma produced by corona discharge that takes place between electrodes. This method is used on flat plastic, for example, films, and occasionally used for irregular shapes.

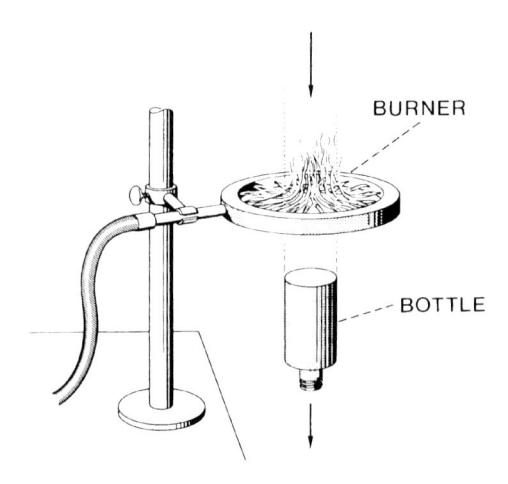

Figure 7.1 A cylindrical item is treated while falling through a ring shaped burner. (Courtesy of Equistar Chemical, LP.)

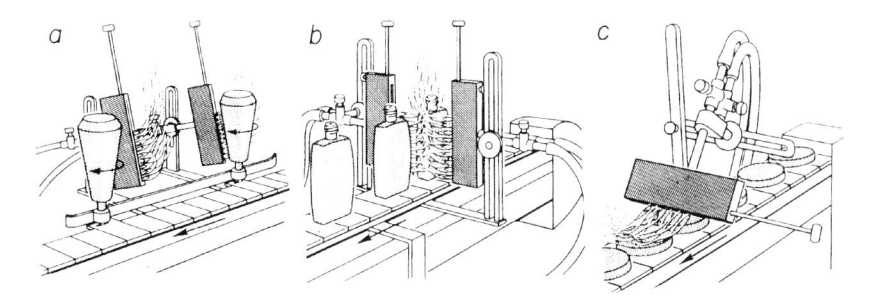

Figure 7.2 This shows how the burners on the machine shown in Fig 7.1 can be adapted to a great variety of shapes and sizes of moldings to be flame treated for printability and adhesion. A, The burners are tilted at an angle to treat cone-shaped bottles. B, Flat bottles are treated on both sides simultaneously by passing burners mounted on opposite sides of the conveyor belt. C, Only one burner is required to treat the top surface of flat moldings. (Courtesy of Equistar Chemical, LP)

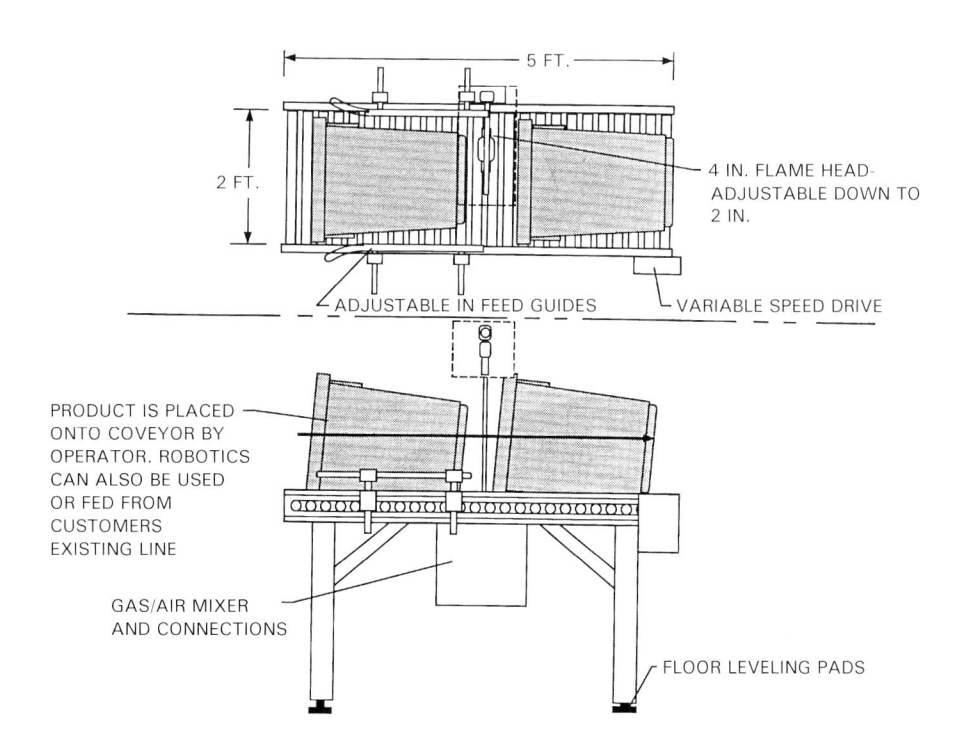

Figure 7.3 Flame treating trash cans

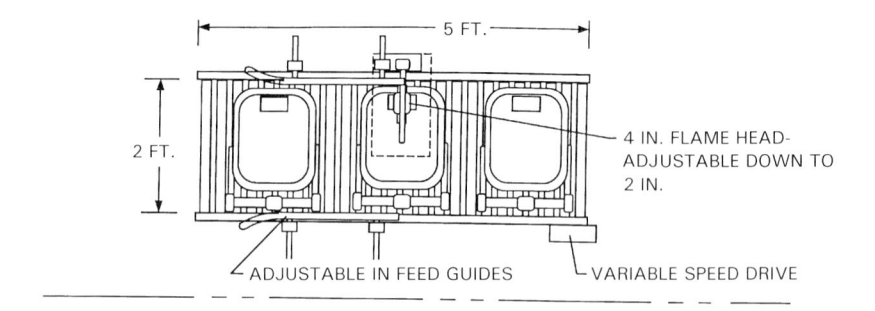

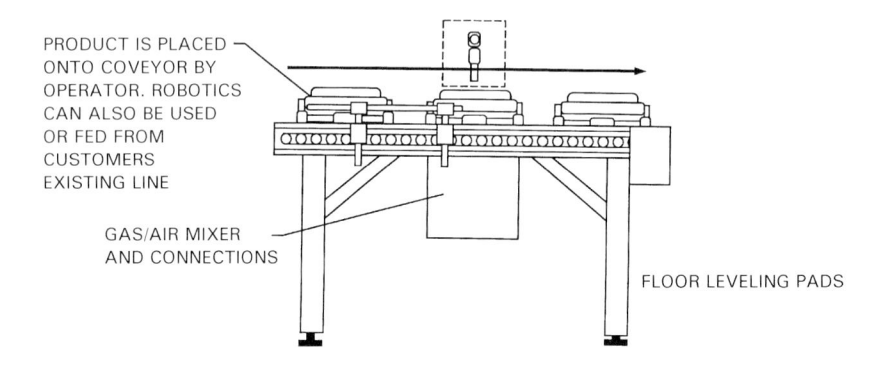

Figure 7.4 Flame treating can lids

7.2.4 Washing with Water-Based Chemicals

Various surface impurities release agents electrostatically or other attracted dust particles and additives may have migrated to the surface. They can be removed with water washing. A cleaning cycle consists of the following steps:

- Cleaner treatment,
- several rinses, and
- finished with a rinse of deionized water [6].

7.2.5 Solvent Cleaning and Etching

Cleaning with solvent can be done by wiping, immersion, spraying, or vapor decreasing. The least effective method is wiping, which often results in distributing the contaminant to the surface, rather than removing it. The better process is immersion, particularly if done with mechanical or ultrasonic scrubbing. It is also much better when followed by a rinse, either by immersion or spray.

Fluorocarbon polymers are usually chemically etched, since they do not respond to other treatments. ABS (Acrylonitrile-Butadiene-Styrene) parts are usually etched chemically for metallic plating.

Etching solutions are oxidizing chemicals and highly corrosive and difficult to handle and dispose.

7.2.6 Additives Compounded into Resins

Plastic material surfaces can be changed with the addition of incompatible additives that migrate to the surface.

7.3 Spray Painting

This decoration method is mainly used for painted coated areas on irregular blow molded articles. Spray painting equipment and processes are discussed in several publications [7, 8].

7.3.1 Air Atomization and Airless Sprays

Air atomization and airless sprays are the most common painting equipment used by blow molders. The operation consists of atomizing a liquid stream by a spray gun. Small paint particles produced by the gun are then delivered to the plastic part surface and remain adhered after the liquid vehicle is dried.

Several atomization techniques are available.

7.3.2 Masking

The spray painting process is excellent for applications involving large areas, irregular surfaces, multiple parts, and separated surfaces. Masking is required to separate areas. By far the simplest method of masking involves employing tape and paper to cover areas of the plastic surface. Masking is used for production of automobile bumpers where high quality is required. The tape must be resistant to the solvents used in the spray paint and be able to withstand the drying conditions, and be separated cleanly on removal. The sharpness along the edge of the tape depends on the tape quality.

Hard masks are reusable, made from metal or plastic, and placed over the object to be painted to shield paint from selected areas on the surface. Among materials used for plastic masks are polyurethanes and silicone rubber. Metal masks are usually either fabricated by standard mechanical methods or electroformed using copper or nickel. These masks are used, and then cleaned repeatedly in a cleansing solvent tank. The masks are usually

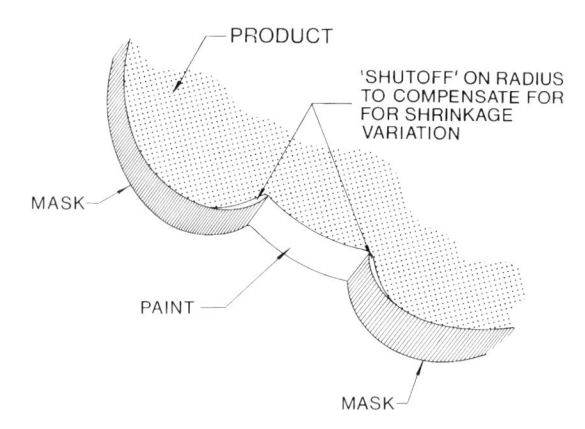

Figure 7.5 Paint spray mask

duplicated or triplicated to allow cleaning while others are in use. A typical spray mask cross-section is shown in Fig. 7.5.

7.3.3 Vapor Degreasing

An effective process is vapor degreasing because it is carried out in a tank with a solvent reservoir on the bottom. When the solvent is heated the vapor condenses onto the cooler plastic surface. The resultant condensate dissolves surface impurities and they are carried away. This process usually takes about 1 minute. The solvents most effective for polyolefins are toluene and those that are chlorinated.

7.3.4 Mechanical Abrasion—Sanding

With mechanical abrasion several process options are available:

- Dry blasting with a nonmetallic grit (flint, silica, aluminum oxide);
- wet abrasive blast (a slurrey of aluminum oxide); and
- hand or machine sanding and scouring with tap water and scouring powder [6].

7.3.5 Chemical Etching

This process entails the exposure of the plastic surface to a solution of reactive chemical compounds. It requires immersion of the part into a bath for a period of time, then rinsing and drying. The resultant treatment is a chemical surface change, oxidation, with an increase in surface wettability (which is the critical surface tension). Also, some surface material may be removed, giving it a micro roughness.

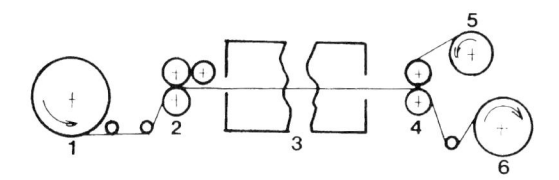

Figure 7.6 Transfer coating of pressure sensitive label stock. (1) Release paper unwind, (2) coater, (3) drier, (4) laminating rolls, (5) label face supply roll, (6) rewind

7.4 Labels

There are three steps to produce labels: production, printing, and die cutting.

The label stock is produced by a transfer coating technique, which entails adhesive being applied to the release lever, dried, and laminated to the face stock, transferring the adhesive from lever to the face stock. Figure 7.6 shows a diagram transferring the coating.

Printing of labels is mostly by a narrow-web method, thus, large label stock rolls are slit to the required width. The printing methods are either flexography or letter press on multicolor narrow-web machines.

7.4.1 Label Application

Large labels are most often applied by hand, while automatic equipment is used for small labels. Labels are secured by wet glue or hot melt adhesive application at the time of label application. However, for blow molded products (usually Polyethylene) pressure-sensitive adhesive preapplied to label stock is more common. When release paper is run over with a sharp edge the label disbonds and continues to move on the straight pass. This is illustrated in Fig. 7.7a.

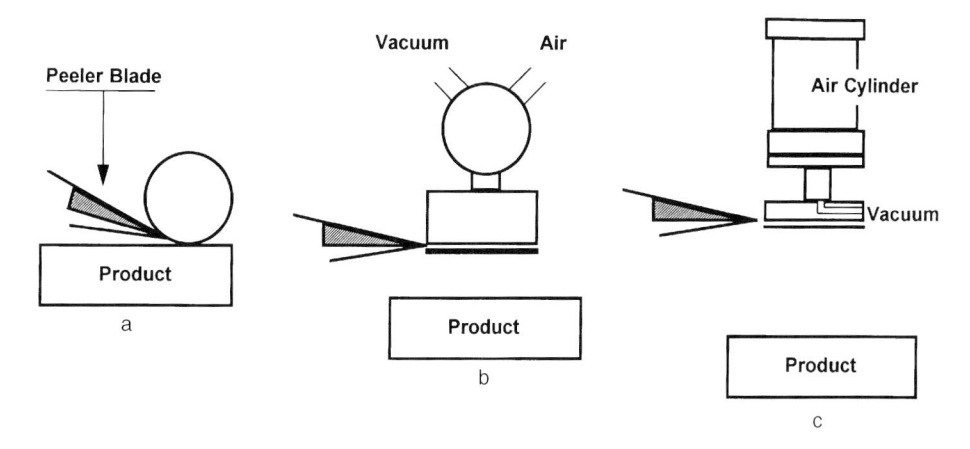

Figure 7.7a Pressure-sensitive label application methods

Large labels have split stock to enable them to be removed easily by hand. (Fig. 7.7b).

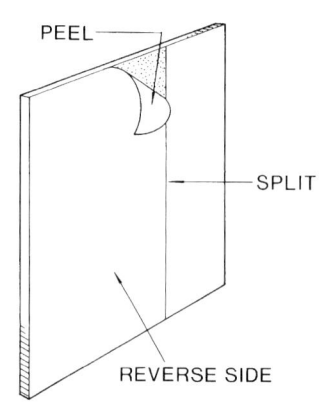

Figure 7.7b Peel-off label

7.5 Screen Printing

Screen printing is a special process in which ink is forced through open areas of the screen. The print pattern is determined by open areas of the screen. This principle is illustrated in Fig. 7.8 [9]. A squeegee forces ink through the screen openings which merge to provide a continuous coating over the desired area (Fig. 7.9). The process allows heavy ink coatings and is not possible in other printing methods. This process is simple and versatile; therefore the equipment is expensive. It is slow, however, and it is not generally used in high production operations [6].

7.5 Screen Printing   93

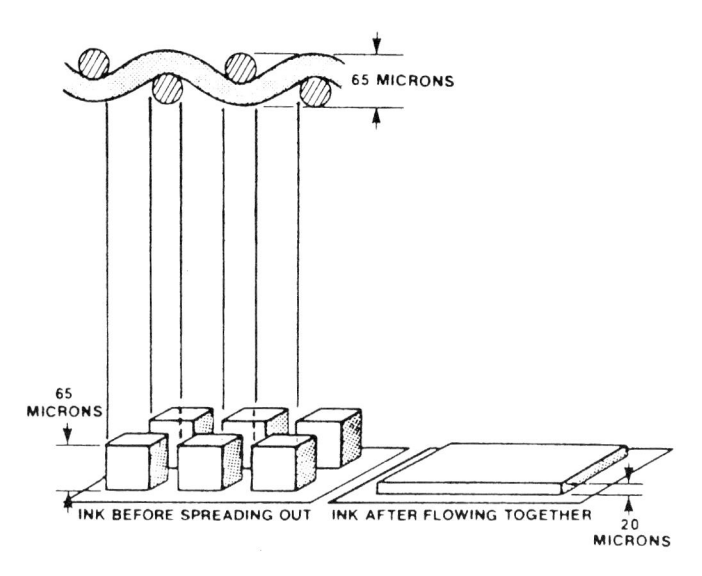

Figure 7.8 Formation of film from ink dots. (Courtesy of Tedko, Inc.)

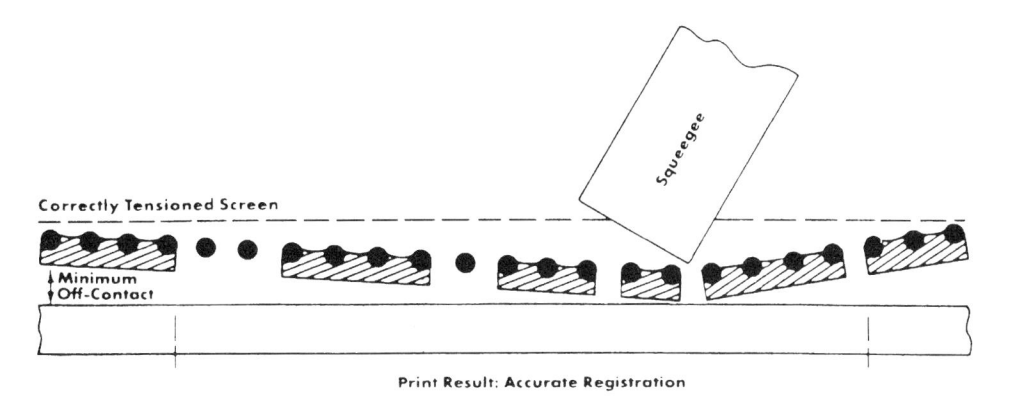

Figure 7.9 Mechanics of screen printing process (Courtesy of Tedko, Inc.)

7.5.1 Screen Printers

Screen printers range from basic hand-operated units to fully automatic machines. They are available in three styles: flat bed, rotary, and cylinder (Fig. 7.10). In the flat bed machine the printing stock and screen are stationary, the squeegee moves along the screen to deposit the ink. Production rates are up to 100 parts per minute.

The rotary screen printer has a drum that is constructed from a metal screen. The squeegee is positioned inside the drum and the ink is pumped into the inside. In this process the squeegee is stationary with both the screen and printing stock moving. The cylinder consists of a stationary squeegee and moving printing stock and screen. The process is similar to the rotary machine except stock is fed intermittently with rates up to 8000 pieces per hour.

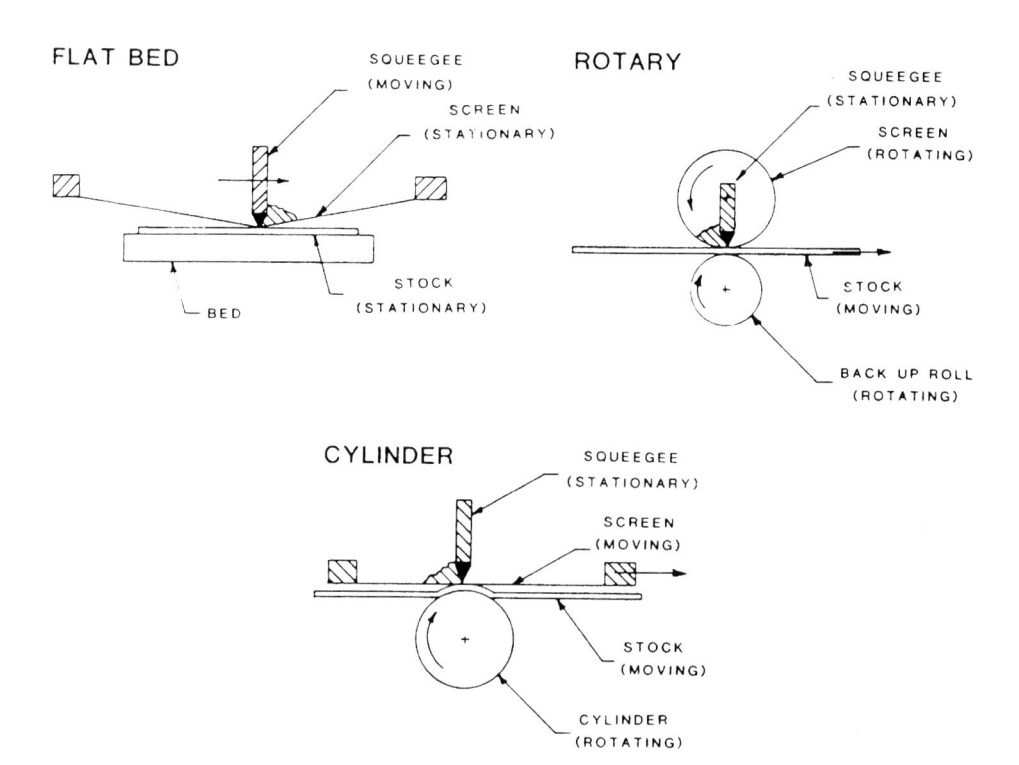

Figure 7.10 Basic types of screen printers (Courtesy of Sheldahl, Inc.)

7.6 Pad Printing

In the pad printing method, the ink is applied in excess over an engraved steel plate (cliché). The excess ink is removed by a doctor blade, so that ink is left only in the engraved recesses. A soft silicone rubber pad picks up the ink from the cliché and transfers it to the part. The stages and basic pad shapes are illustrated in Figs. 7.11 and 7.12.

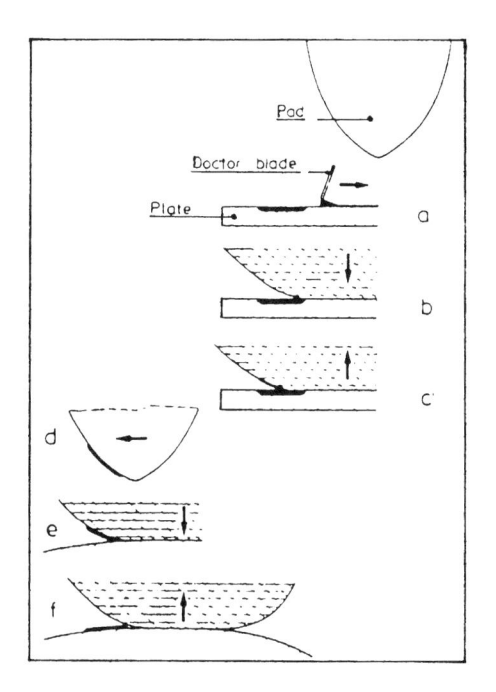

Figure 7.11 Stages of pad printing: (a) Doctoring to remove excess ink. (b) Initial contract with cliché (plate). (c) Ink removal from cliché. (d) Pad moving to printing surface. (e) Initial contact of printing surface. (f) Removal of the pad. (Courtesy of Midwest Technical Service)

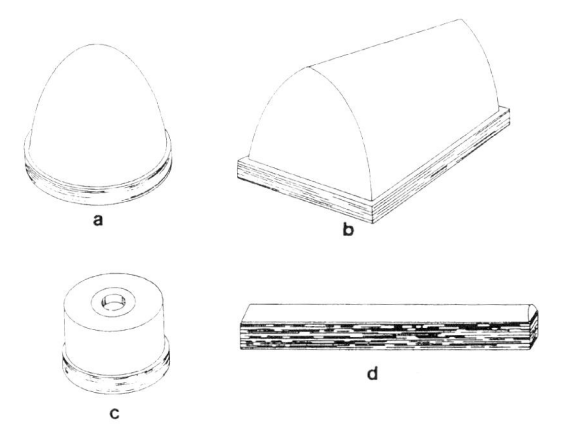

Figure 7.12 Basic pad shapes: (a) Standard. (b) Wedge. (c) Doughnut. (d) Ribbon (Courtesy of Midwest Technical Service)

7.6.1 Pad Equipment

Pad printing is predominately a reciprocating process, although a rotary version was introduced recently. The reciprocating process is slow, as it operates at a production rate of 30 to 40 cycles per minute; however multiheads may be used (Fig. 7.13). This equipment is more expensive than that for stamping or screen printing.

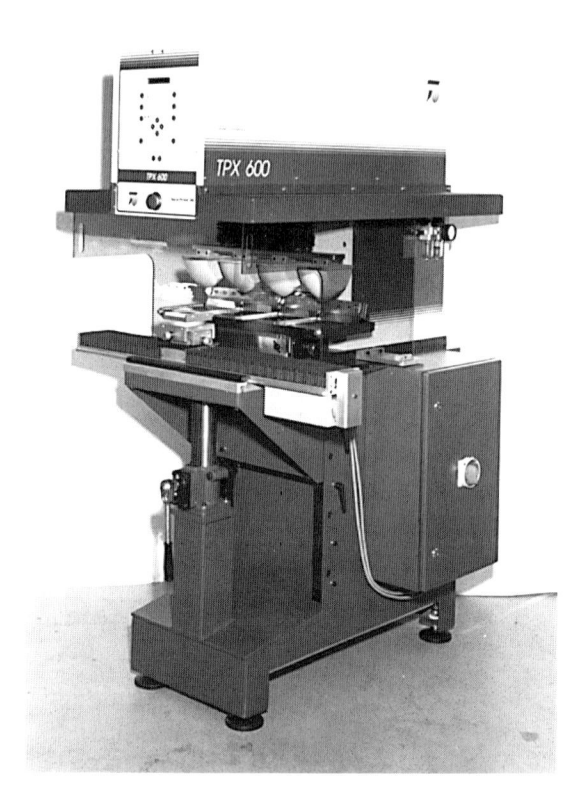

Figure 7.13 Four station pad printer (Courtesy of Teca-Print USA, Billeria, MA)

7.7 Hot Stamping

This process involves the transfer of pigments from a carrier to the part to be printed by application of heat and pressure. Hot stamping is performed by bringing a heated die in contact with the product to be decorated. A stamping foil is placed between the two and a mirror of the engraved pattern on the die is transferred to the part. A flat die can be used to transfer to high points on the surface of the part, for example, lettering. It should be noted that in blow molding large industrial parts the wall thickness is often uneven, thus, the impression is in effect branded into the part to ensure complete coverage. This impression should be kept simple for any large area. (Fig. 7.14).

Figure 7.14 Hot stamp examples on municipal roll out refuse carts (Courtesy of ZARN, Inc., Reidsville, NC)

98 Decoration of Blow Molded Products

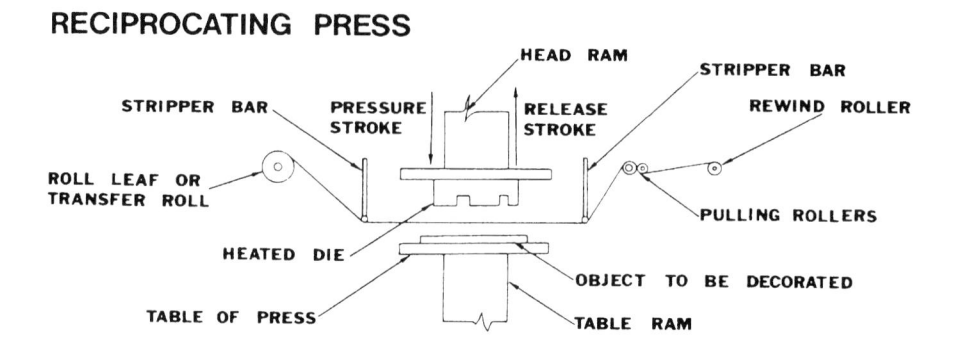

Rotary presses with cylindrical and flat dies are also shown in Fig. 7.15.

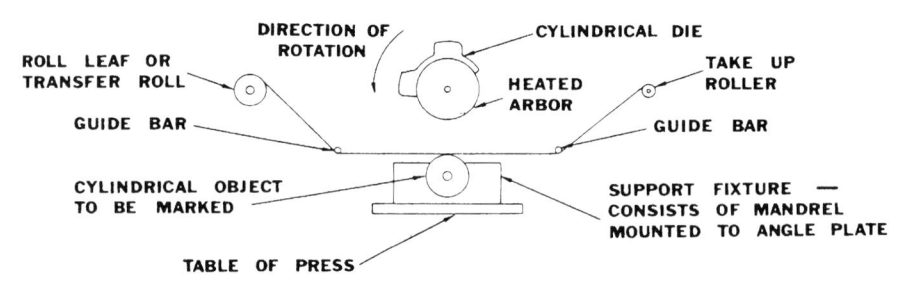

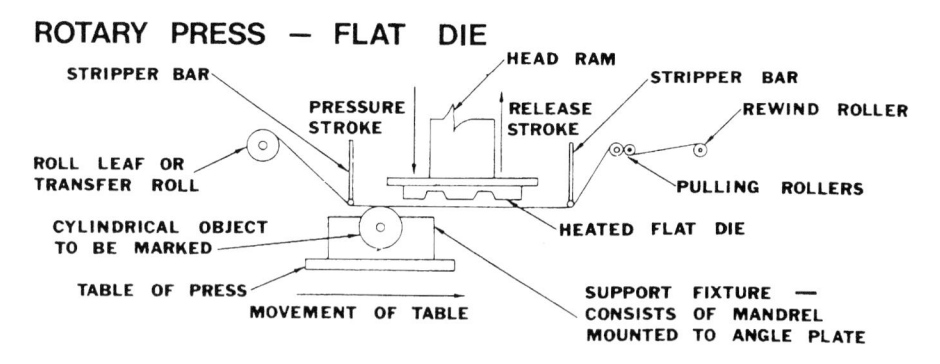

Figure 7.15 Hot stamping processes (Courtesy of Kensol-Olsenmark)

7.7.1 Hot Stamping Foils

Hot stamping foils, sometimes referred to as roll leaf, consist of several layers:

- Film carrier,
- release coating,
- decorating coating, and
- adhesive coating.

Polyester film (0.25 mm or 0.001 in. thick) often is the carrier of choice since it has sufficient strength and temperature resistance to resist transfer operation.

The release coating consists of wax or a resin with a melting point such that it liquefies when contacted by a hot die. This provides for the separation of the film from the decorative coating.

The decorative coating is pigmented to provide the image on the part. It often consists of several coats. A clear top coating is used over the metallic coating so the color shows through.

7.8 Decals

The decal images are transferred by heat and pressure using conventional or specially constructed hot stamp presses, the type used for foil marking. This permanently welds the image to the receiving surface (Fig. 7.16).

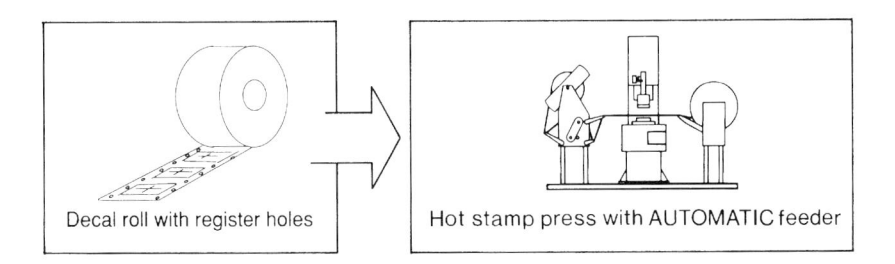

Figure 7.16 Decal hot stamp press

7.8.1 Advantages of Heat Transfers

- Equal, or superior, image quality is possible at lower cost than direct printing.
- No wet ink, solvent, or cleanup problems. Saves energy needed for drying ovens.
- Product can be handled immediately after image application.

- Process is more "forgiving" on uneven surfaces.
- Application can often be coordinated with another operation—plastic molding—to save time and labor.
- Unskilled operators can operate decal equipment.
- Application possible on finished parts to cut inventory expense.

7.9 In Mold Labeling

If the bottle or small part is under 6 liters (5 gallons) and the annual volume is 2 to 8 million units per year then in mold labeling (IML) should be considered. The initial investment is high and set up times much longer but with the higher volumes IML is competitive in price with post-mold labeling/decorative Systems [10].

7.9.1 In Molding Labeling Equipment

Setup of and maintaining IML systems has become easier in recent years because several machine manufacturers have developed molding machines that are better adapted to mounting labeling hardware and they offer complete machine, molds, and labeling hardware packages. For example, Battenfield produces a long stroke models that have more room for mounting IML equipment than the more commonly used swing arm machines or shuttle machines. Krupp Kautex also has a long stroke KBS-1 family of machines. Graham Engineering has introduced a novel type long stroke machines. The techne Graham Model 6000 twin has dual clamps side by side on a single carriage. Bekum has considerable success with IML in its B.M.-704 swing arm machine series (Fig. 7.17) [10].

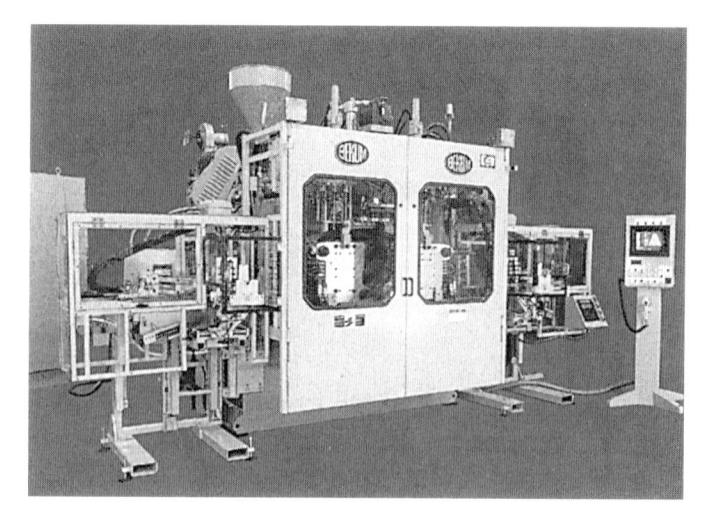

Figure 7.17 IML equipment by Avery Dennison/Autotec flanks the sides of this Bekum H-121 dual-sided shuttle blow molder (Courtesy of Bekum)

7.9.2 The In Molding Labeling Process

The basic IML process is suitable for rectangular and oval containers, and usually employs paper labels produced from medium base stock coated on one side for printability and the other for heat seal coating. Die cut labels are placed in magazines from which robotic arms pick them up and insert them into the mold. Plastic labels have eliminated the bulge experienced with paper labels, and act as insulators and decrease the cooling rate of the plastic area covered by the label. A further advantage of plastic labels is that they are directly recyclable to the blow molding process. In mold labels are printed by any one of the common printing methods [11].

7.9.3 "In Mold" Label Molds

Molds that have been modified or designed for IML are fitted to both vacuum position and retain the "in mold" labels. Owing to the negative air pressure effect (vacuum), such molds will usually have increased capacities proportional to the shrinkage reduction obtained resulting from the vacuum. When volumetric filling of such containers is employed, compensating adjustments in filling line height must be made [12]. For example, a cubical container 152 mm ($6 \times 6 \times 6$ in.) would require a 0.24 mm/1 cm (0.024 in./in.) shrink allowance, and a blow mold cavity of 156 mm (6.144×6.144). If vacuum assist is exerted on the mold in conjunction with normal blow pressures of 60 to 90 psi (4.137 to 6.2055 bars) during blow molding the effect would be to produce a container having an outside finished dimension 154 mm (6.048×6.098) or an increase of 0.086 liter (5.23 in.^{3}). Such an increase may result as a costly giveaway product [12].

Figure 7.18 In mold labeling operation (Courtesy of Bekum)

7.9.4 Cycle Times

Cycle times can be affected in two ways: The mechanical delay and the process delay.

Mechanical delays are related to the speed at which the robot arm or injection device can place the label in the mold and then retract from the mold area.

The process delay results from additional time required to cool the areas of the bottle that are covered by the label and thus do not come in direct contact with the mold. The room temperature label comes in contact with the hot parison and the label becomes an insulator [10].

7.9.5 Aesthetics

IML has inherent aesthetic benefits. The bottle is molded around the label and encapsulates the label edges, providing a smooth, seamless surface, sometimes referred to as a ''no label'' screen printed look.

Transparent labels are much more common than paper in IML though they are approximately 50% more expensive. They give a glossier appearance and thereby better simulate the look of printing. Heat activation is also used on plastic labels which improves the bonding to the plastic bottle. The result is that in mold film labels do not peel, curl, or blister like paper labels on shampoo bottles and other personal care products where they are subject to heat and moisture exposure.

A cleaner looking package results from IML and a vibrance of ten colors compared with screen printing, which is usually limited to six colors [10].

7.10 Conclusion

There are many different options for decorating plastic blow molding products: labels, decals, screen printing, hot stamping, pad printing, and paint. Each process has its own appeal to the customer and is specialized; therefore each cannot be covered in detail in this book.

7.11 References

1. Satas, D. (Ed.) *Plastic Finishing and Decoration*. (1986) Van Nostrand Reinhold, New York
2. Margolis, J.M. (Ed.) *Decorating Plastics* (1986) Hanser, Munich
3. Stoeckhert, K. (Ed.) *Veredeln Von Kunstoff-Oberflaechen* (1974) Hanser, Munich
4. *Modern Plastic Encyclopedia* (1988) Mcgraw-Hill, New York
5. Stats, D. *Decorating Plastics.* In *SPI Engineering Handbook*. Berins, M.K. (Ed.) (1990) Van Nostrand Reinhold, New York
6. Lee, N.C. (Ed.) *Plastic Blow Molding Handbook* (1990) Van Nostrand Reinhold, New York
7. Van Hor, R.C. Spray Painting Plastics. In *Plastic Finishing and Decoration*. Statas, D. (Ed.) (1996) Van Nostrand Reinhold, New York
8. Statas, D. *Spray Coating*
9. Gillzo, K.B. Screen Printing. In *Plastic Finishing and Decoration*. Statas, D. (Ed.) (1986) Van Nostrand Reinhold, New York
10. Knights, M. Is In-Mold Labeling in Your Cards? *Plastic Technology*, May 1995
11. Lee, N.C. (Ed.) *Plastic Blow Molding Handbook. D. Status Decoration Equipment and Processes, In Mold Labeling* (1990) Van Nostrand Reinhold
12. Rosato, D.V., Rosato, D.V. (Eds.), *Blow Molding Handbook*, Hanser, Munich

8
The Blow Molding Process

As indicated earlier, three major processes (extrusion, injection, and stretch [1]) are used to produce the parison or preform from which a part may be formed by blow molding. In this chapter the processes and machines are explained in further detail.

8.1 Extrusion Blow Molding

As the name implies, the machine has an extruder. When thinking about a basic extruder what may come to mind is the old simple meat mincing machine which was found in many kitchens (Fig. 8.1). Chunks of meat are fed into the funnel (hopper) and an arbor, fluted

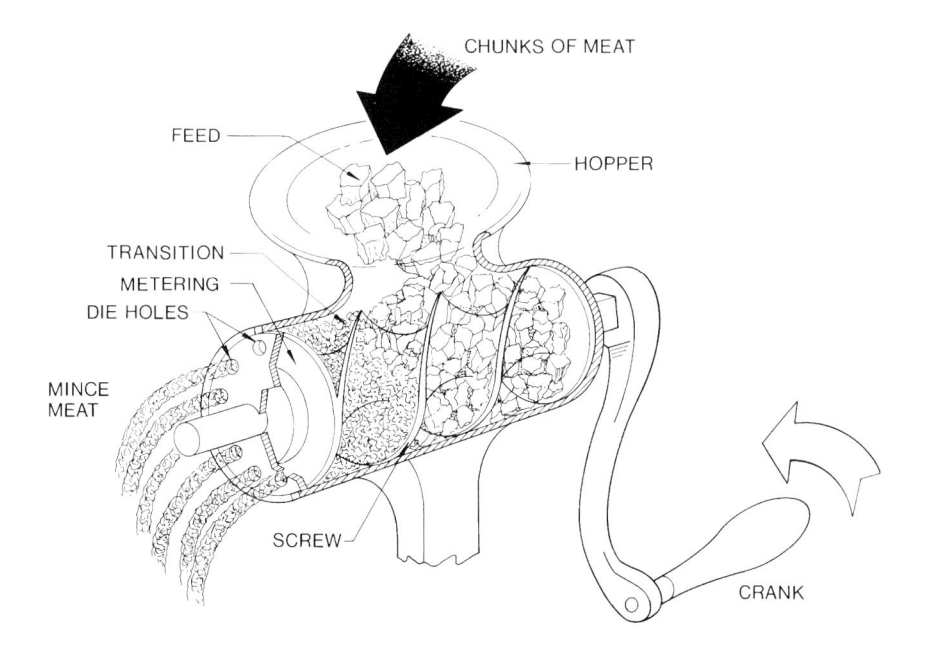

Figure 8.1 Old kitchen mincer

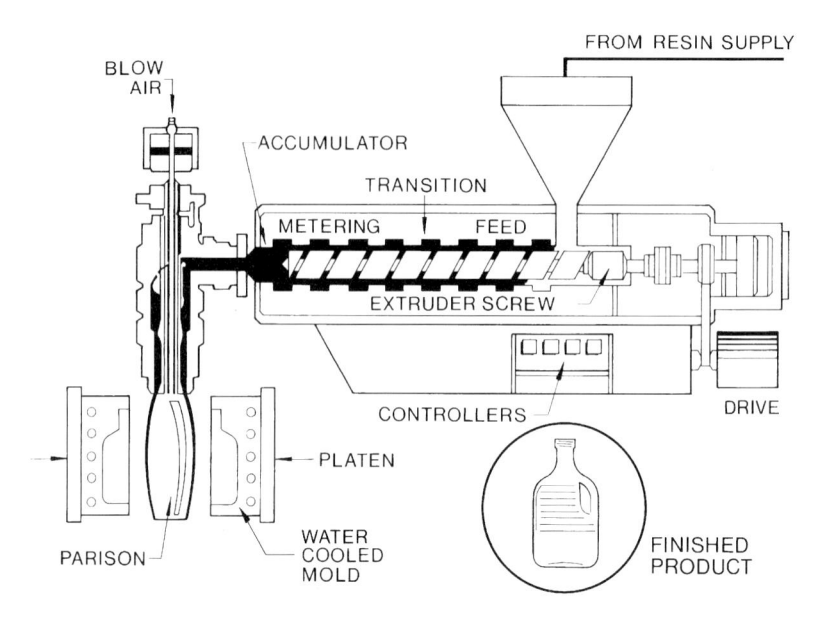

Figure 8.2 Reciprocating screw machine

screw moves meat (material) forward, by the turning of a crank, through a die at the end, producing mince meat. The plastic extruder machine is far more complex but operates much the same. Plastic resin pellets are fed into a hopper and a fluted screw is turned by a variable speed drive with an electrical motor, or hydraulic drive, which propels the material forward (Fig. 8.2). The barrel has heater bands for melting the plastic material. The screw normally has three sections: feed, transition, and metering, which process the plastic material through a head which contains a die and mandrel for forming the parison. In addition to this basic, single-screw type extrusion machine, twin-screw type extruders are occasionally used in blow molding.

8.1.1 Understanding the Extruder

Knowing and understanding the extrusion process is important when we consider that 65% of all resin is processed by the extrusion method. The most widely used extruder is the single-screw machine. By examining the components of this machine a working knowledge of its operation can be obtained.

The most important element in the extruder is the screw. The screw is responsible for plasticating or melting and mixing the polymer to a homogeneous state. Figure 8.3 shows that the extrusion screw is divided into three sections: the feed, transition, and metering.

The feed section of the screw receives the solid pellets from the hopper and conveys them forward. The feed section is always designed with flight depths that are greater than

8.1 Extrusion Blow Molding

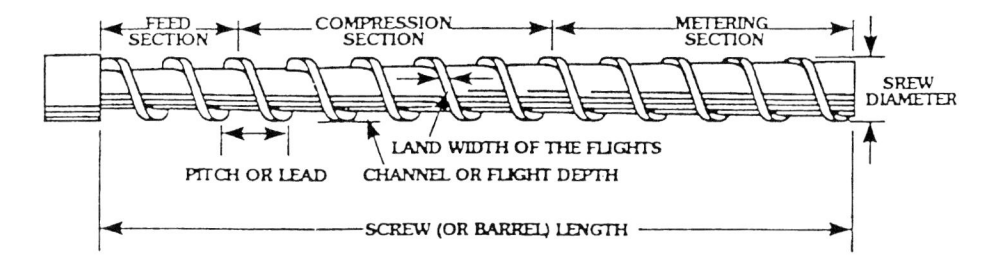

Figure 8.3 Polyethylene screw (Courtesy of Equistar Chemicals, LP.)

those of the other sections. This assures that a steady flow of material is always conveyed into the transition section. Little or no melting of the plastic takes place in the feed section.

The transition section has a progressively changing root diameter. This feature compensates for the change in volume as the polymer melts and maintains the pressure on the pellets, which in turn increases the frictional or shear heat generated. The temperature in this zone must be monitored closely. The shear generated can often raise actual temperatures 10 to 50 degrees above the set point. Many materials experience 90% of their melt in this zone. Most of the mixture of resin with color and other additives occurs in this zone.

The metering section also has a constant root diameter, but with the least flight depth of any section of the screw. This section provides the final mixing to a melt of uniform temperature, and acts as a pump to the accumulator. Flight depth usually depends on the screw size and is a complex function of screw diameter.

Extruder screws are classified by their length-to-diameter ratio or L/D. The length is the measurement of the flighted portion of the screw. The diameter is the OD of the flights. The most common ratios are 24:1 to 30:1. The higher the ratio the higher the surface area available for shearing, mixing, and plasticating.

To customize the extrusion process to certain groups of engineered resins, new screw designs have been developed. Most of the new designs have had changes incorporated into the metering sections. The object is to improve uniformity of the melt without major increases to the material's temperature.

The spiral, barrier type screw is commonly used in the industry (Fig. 8.4). The barrier screw has a sectioned flight that has a reduced diameter to allow the melt to flow into its own channel and maintain the melt/solid separation [1].

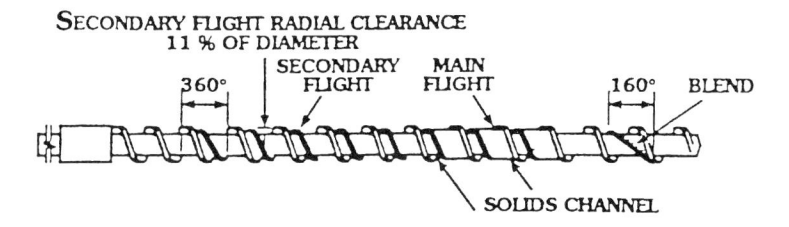

Figure 8.4 Barrier screw (Courtesy of Equistar Chemicals, LP.)

8.1.2 Blow Molding Technique

The blow molding process generally consists of three stages:

1. Melting, or plasticizing the resin,
2. forming the parison or preform, and
3. inflating or blowing the parison in the mold to produce the end product.

Figure 8.5 is a schematic drawing of the steps that comprise the three stages, the actual blowing stages, or molding cycle.

However there are four steps of the actual blow molding cycle, plus a fifth step for trimming (1, 2, 3, 4, and 5 in Fig. 8.5).

1. Melting, or plasticizing, the resin
2. The molten, hollow tube—the parison or preform—is placed between the two halves of the blowing mold.
3. The blowing mold closes around the parison.
4. The parison, still molten, is pinched off and inflated by an air blast which forces the molten parison against the inside contours of the cooled mold. When the piece has cooled enough, the blowing mold is opened and the formed piece is ejected.
5. Trim excess flash from the part, which takes place during stage 2, unless detached in the mold.

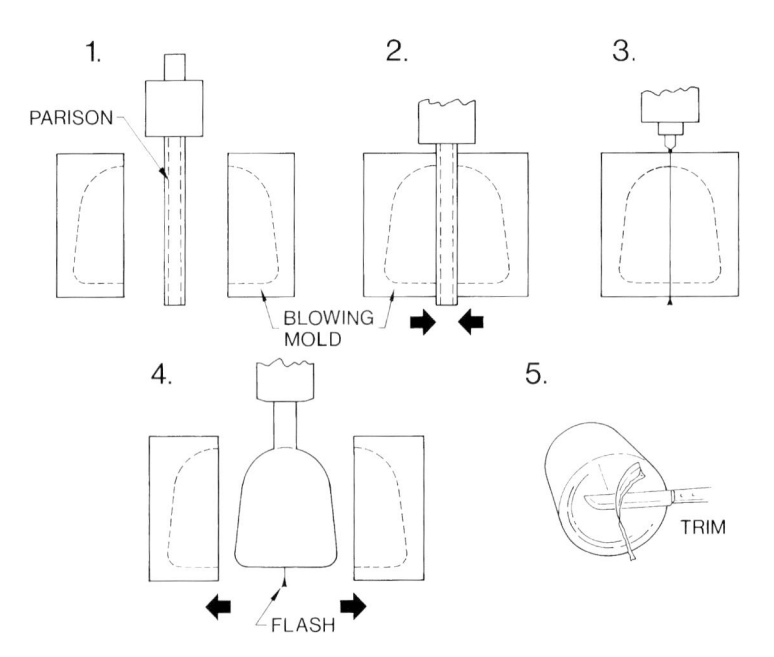

Figure 8.5 Mold cycle sequence

8.1 Extrusion Blow Molding

To make the parison, the plastic pellets or cubes first have to be melted, plasticized, and well mixed under heat and pressure, and then shaped into a tube. Several pieces can be produced simultaneously on one machine, with the above steps overlapping. Blowing takes a certain amount of time, measured in seconds or, for large items, a minute or more. Most of the molding cycle is taken up by the cooling step and therefore it is the cooling time that controls the blowing cycle. Speed of the machine that melts the resin and produces the parison may conform with the blowing cycle, especially cooling time. The fifth step, trimming excess flash, takes place after part ejection.

Flash is a characteristic of all extrusion blow molding. It is formed when the parison is pinched together when captured by the mold. With most resins it can easily be reclaimed.

In many cases the entire sequence can be totally automated and directly connected to other downstream operations, such as container labeling and filling.

The parison, or preform, is a circular tube, sometimes elliptical (see parison shaping), consisting of resin in the molten state which permits it to be blown into its final shape (stage 3).

Figure 8.6 shows a typical blow molding cycle for 1-gallon (3.875 liters) containers. Three phases of the cycle are noted and each phase is broken down into its components. All blow-molding cycles contain these segments although the time intervals may vary somewhat.

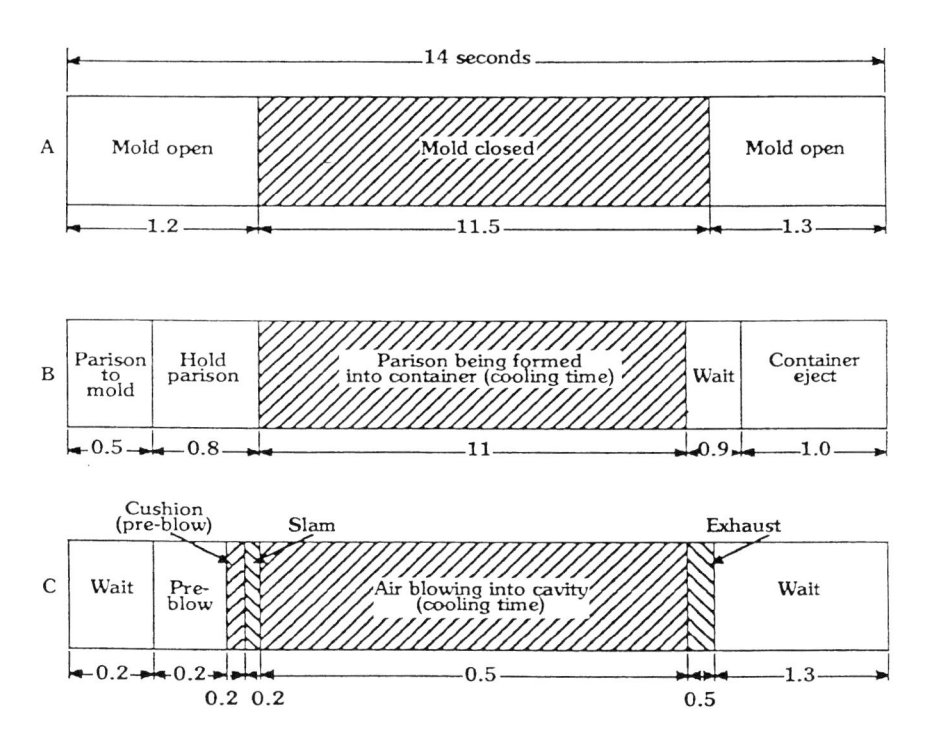

Figure 8.6 Blow molding cycles (Courtesy of Equistar Chemicalsw, LP.)

8.1.3 Continuous Extrusion

Continuous extrusion is one of two basic approaches to blow molding, the other being intermittent. In continuous extrusion the parison is formed continuously as the article is molded, cooled, and removed. To avoid interference with parison formation, the mold clamp mechanism must be moved to a blowing station; in doing this several methods are available. They generally fall into three categories: shuttle, rotary wheel system, and rising mold. To avoid interference with the parison formation, the mold clamping mechanism must move rapidly to capture the parison and return it to the blowing station where the blow pins enter. It is used for all commonly blow molded plastic resins but is best suited for (poly)vinyl chloride (PVC) and other heat-sensitive resins. Because of the gentle uninterrupted flow, continuous extrusion reduces the chance of thermal degradation (Fig. 8.7).

In this process the parison is continuously being formed. The rate of formation is synchronized by a variable speed control to the rate of part blow molding, cooling, and removal. Normally continuous extrusion equipment is used for containers up to 3.875 liters (1 gallon) in size.

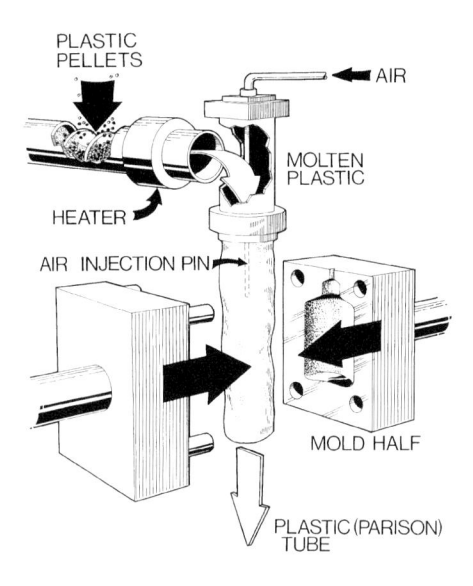

Figure 8.7 Continuous extrusion

8.1.3.1 Shuttle System

With the shuttle system, the blowing station is located on one or both sides of the extruder, as in Fig. 8.8. Once the parison reaches the proper length the clamp mechanism quickly shuttles from the blowing station to a position under the extrusion die head, captures and cuts the parison, and then quickly returns to the blowing station. Clearance is thus provided for the next parison. With this style of continuous extrusion equipment, multiparison heads are often used to increase production yield (Fig. 8.9).

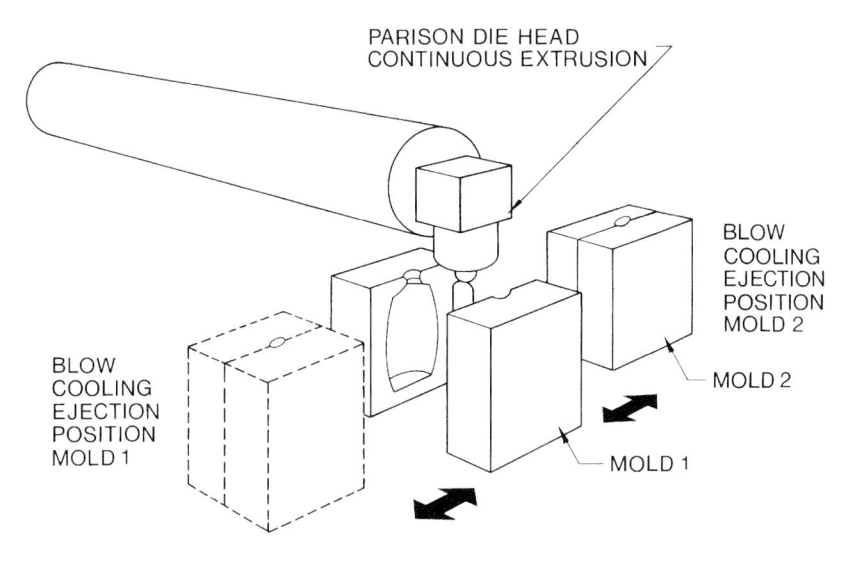

Figure 8.8 Continuous extrusion with shuttle mold system

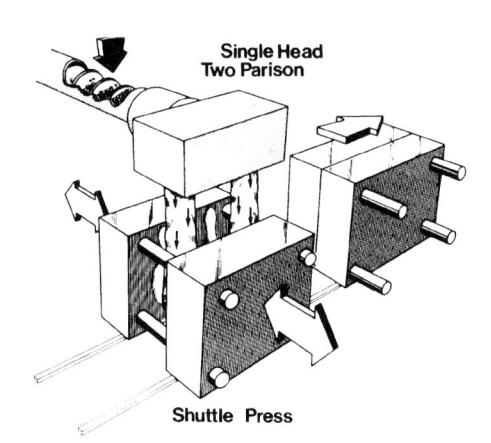

Figure 8.9 Continuous extrusion with multi-heads and shuttle system

8.1.3.2 Rising Mold

The parison is continuously extruded directly above the cavity. At the proper length the mold rises to capture the parison and returns downward to the blow station. After the article is blown, the mold opens and the part is removed and the process is repeated.

8.1.3.3 Rotary Wheel

With the rotary wheel system, up to 12 clamping stations are mounted to either a horizontal or vertical wheel. The wheel is rotated past a single or dual extrusion head where the parison is captured, as in Fig. 8.10. At any given moment, the parison is being captured, a

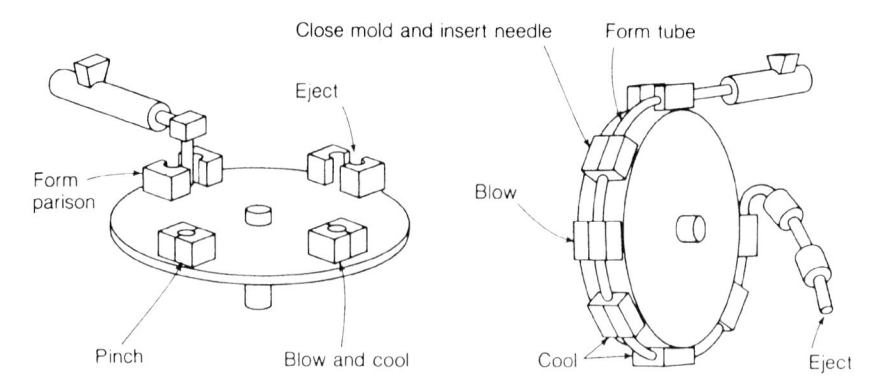

Figure 8.10 Continuous extrusion with either horizontal or vertical wheels

part is being molded, a molded part is being cooled, and a cooled part is being removed. Rotary machines can provide high production yields, but one disadvantage is the complexity and setup of the clamp mechanism. They are usually not suited for short production runs. Setup is often done on a spare wheel off-line and then exchanged.

8.1.4 Intermittent Extrusion

Intermittent extrusion is best suited for polyolefin and other non-heat-sensitive resins, and permits the use of a very simple and rugged clamp and part removal system.

In this process the parison is quickly formed immediately after a part is removed from the mold. The mold clamp mechanism does not need to be transferred to a blowing station. Blow molding, cooling, and part removal all take place under the extrusion die head.

Intermittent extrusion machines fall into three general categories: reciprocating screw, ram accumulator, and accumulator head systems. In all three, multiparison heads are often used (see Fig. 8.11 for typical die head).

8.1.4.1 Reciprocating Screw

With the reciprocating screw extrusion system, after the parison is extruded, melt is accumulated in front of the screw as it moves backward. After the previously molded part has cooled, the mold opens, the part is removed, and immediately the screw is rammed forward by hydraulic pressure, forcing the melt through the die head to quickly form the parison. The amount and rate of accumulation is synchronized with the size and mold cooling rate of the part (Fig. 8.2). Normally reciprocating screw machines are used for containers up to 9.46 L ($2\frac{1}{2}$ gallons) in size. Some machines for small containers can extrude as many as 12 parisons at a time.

8.1 Extrusion Blow Molding   111

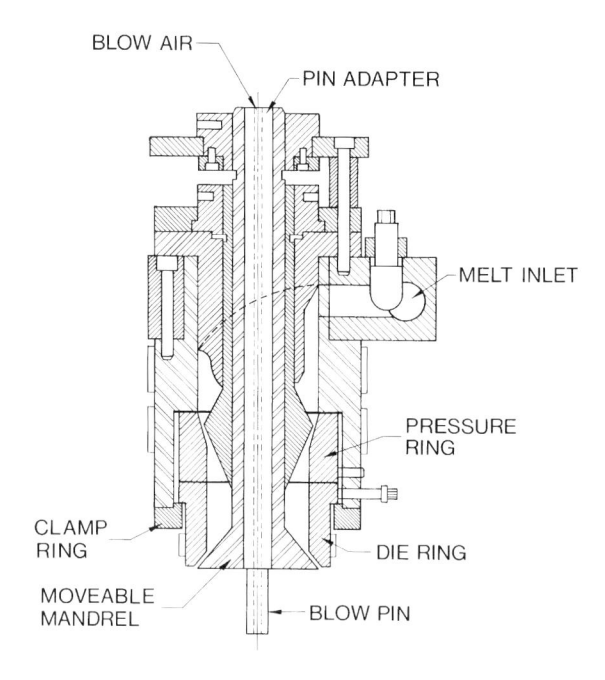

Figure 8.11 Typical die head for intermittent parison extrusion

8.1.4.2 Ram

With the ram accumulator system, melt is accumulated in an auxiliary ram cylinder, usually mounted alongside the extruder. The system functions exactly as a reciprocating screw. It is used to rapidly extrude heavyweight parisons where parison drawdown or sag must be minimized. The system has a disadvantage in that the melt that first enters the accumulator is the last to leave. As a result the melt history of the resin is not uniform. The system normally is used for parts weighing 2.26–45.35 Kg (5 to 100 lbs) (Fig. 8.12). Note that some larger machines are available in Europe. This method is uncommon and not many of these systems are in use at this time.

8.1.4.3 Accumulator

The accumulator head system has the better characteristics of the two previously described systems. It serves as an extruding die head, filled directly by the extruding die, and as an accumulator with first-in/first-out melt flow path.

In a typical accumulator head (Fig. 8.13), the material enters as the outer die body and moves upward to a predetermined shot size limit and then moves downward to rapidly extrude the parison. Used for heavy parts, the system can provide lower molecular stress in the parison. The accumulator directly extrudes the parison from the head annuals with a low uniform pressure. This reduces the excessive pressure drops often found in manifold and head entry areas of other systems.

112 The Blow Molding Process

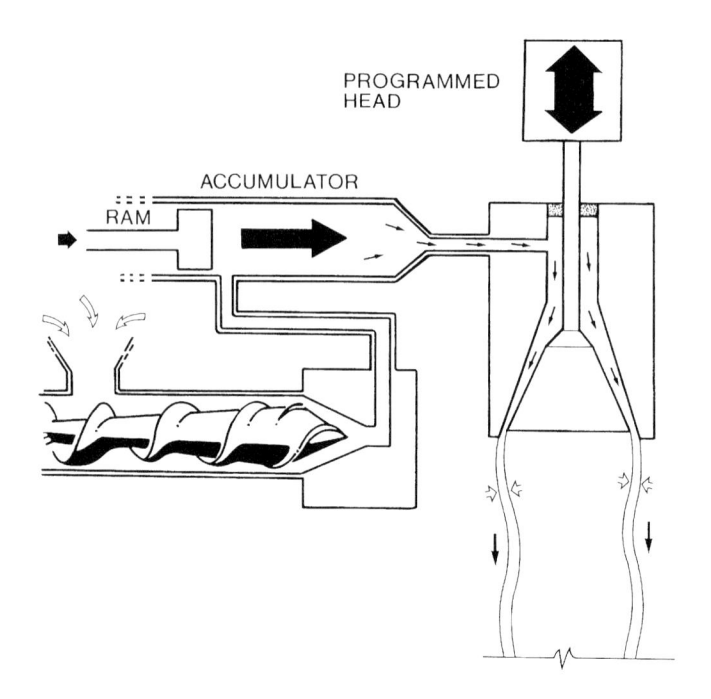

Figure 8.12 Accumulator process

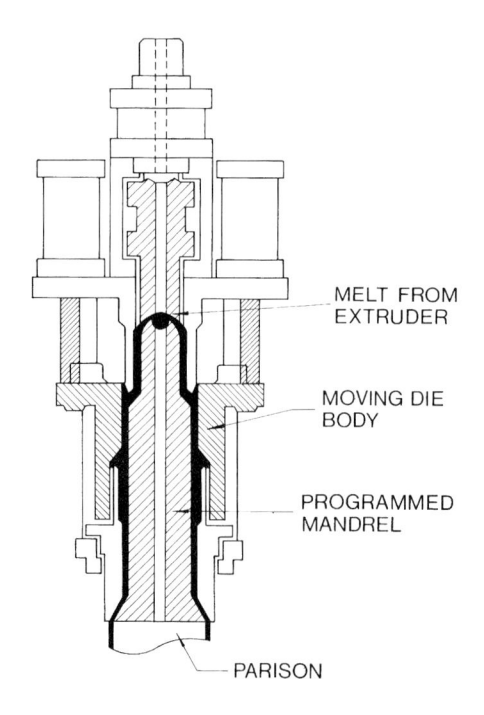

Figure 8.13 Accumulator head

8.1.5 Coextrusion

Coextrusion blow molding is based on the extrusion of a parison made of two or more resin layers. The process offers the advantage of combining dissimilar materials with specific qualities.

For example, a fuel tank can be made of five layers: high density polyethylene (HDPE)/adhesive/nylon/adhesive/HDPE. The nylon provides substantial resistance to gasoline vapor permeation, and the adhesive on either side ensures bonding to the PE. A squeezable catsup bottle can be made of five layers: HDPE/adhesive/ethylenevinyl alcohol/adhesive/HDPE. The ethylenevinyl alcohol (EVAL) copolymer provides additional resistance to oxygen permeation.

Both continuous extrusion and accumulator head extrusion systems have been used. In both cases an individual extruder is used for each resin/layer, as in Fig. 8.14. The process has the disadvantage in that the scrap and flash from molding is not easily reclaimed.

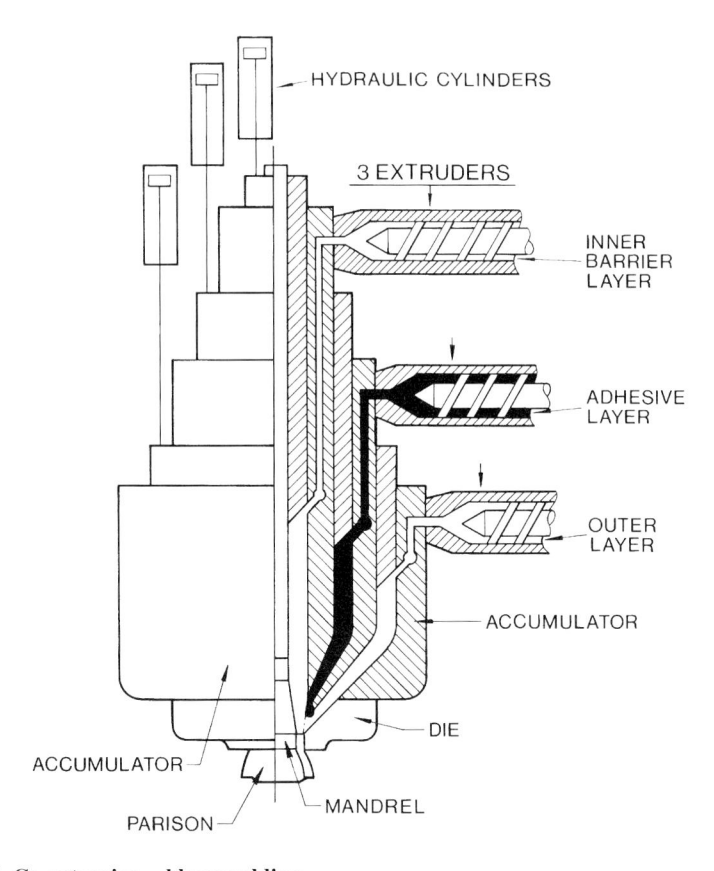

Figure 8.14 Co-extrusion—blow molding

8.1.6 Introduction to Head Tooling

The primary components of head tooling are pin adapter, mandrel, and die and the secondary components are pressure ring and clamping ring (Fig. 8.11).

The pin adapter is that part that attaches to the very center of the head. The tooling adapter is threaded on the top end and threads into the programming rod (the movable part).

The mandrel pin is the part that determines the parison diameter and that bolts to the pin adapter and is bell shaped (diverging) or pointed (converging).

The die is that part that surrounds the tooling pin and has an opening through which the pin protrudes. The opening shape has nearly the same (usually 2 degree difference) angular shape as the tooling pin. As the tooling moves downward the gap increases, allowing a greater wall thickness of the parison to develop as it is pushed out by the ram. As the tooling follows the parison profile, by opening and closing of this gap the wall thickness will vary. The die may be either one or two pieces. In the one-piece design the die clamping ring must be removed and the entire die changed when utilizing a different size tooling; the two-piece design requires removal of only that section that conforms to the tooling size and shape. The new size pin and tooling can then be installed.

The clamping ring is a circular part that fits over the die (one piece) or die adapter (two piece). The clamping ring has a row of bolts that attach it to the head body. The ring and the die (die adapter) have a shoulder to hold the die against the sealing ring. The clamp ring also has (four) bolts around its outer circumference which are used for equalizing the wall thickness which allows the parison to drop straight.

The sealing ring (pressure ring) is a sharply tapered ring that fits into the head body above the die. It serves several purposes, primarily to form a seal between the plastic in the head and any gap or imperfection between the die and head body. It also creates a longer land over which the plastic flows to minimize die swell of the plastic and possible fracturing of the melt. The back pressure created by the seal ring forces some plastic upward past the ram out through the weep hole; thus preventing material degradation above the ram which could limit shot size.

The die and mandrel assembly in the head is usually the final factor in determining the size and shape of the parison (Fig. 8.15).

Choosing the proper head tooling is important if a part is to be molded successfully. To make this choice it is necessary to know the elements of each tooling type.

8.1.6.1 Converging

Converging tooling can be recognized by its shape. The tool is conical, with the small diameter pointing downward when installed. This tooling type must move up into the accumulator to allow material to exit the head during that portion of the cycle where the ram pushes the material out following a programmed profile. Converging tooling exhibits greater material swell owing to the memory of the plastic and the land angle. This tooling is often used where parison sizes are 12.7 cm (5 in.) in diameter or less. Proper tooling calibration is important to prevent damage to the die busing. Maintaining adequate tooling temperatures is easier with this type because of its lower material mass.

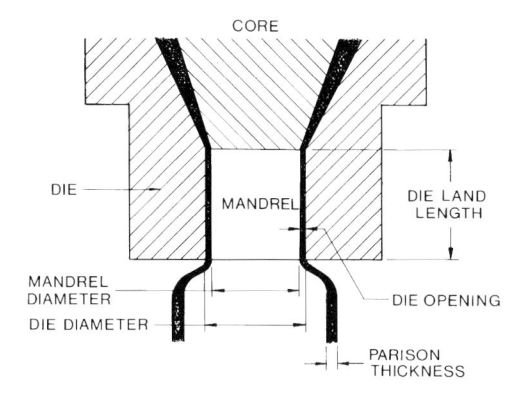

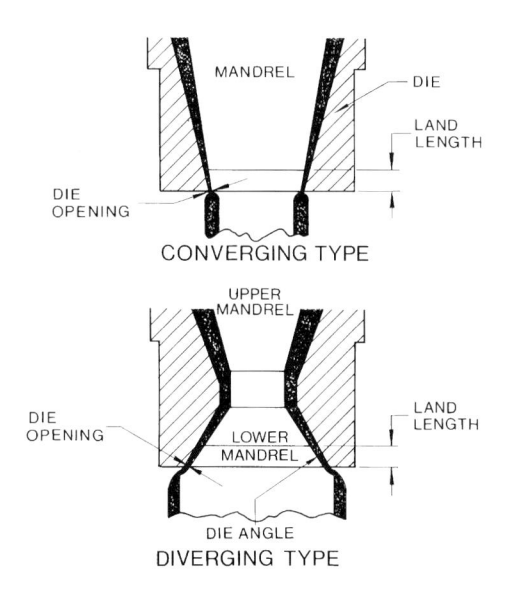

Figure 8.15 Die and mandrel assembly

8.1.6.2 Diverging

This tooling can be recognized by its bell shape. Unlike the converging tool, this tool must move downward to allow material to exit the head. Diverging is typically used to form parts that require parisons above (6 in.) in diameter. The increased size and angle associated with this tooling may require increases in hydraulic pressure to support it. The size of the tool may require core or internal heaters to maintain the metal at its desired temperatures.

The extended land length of the die helps to provide slight support to the parison while it is being clamped upon. It is usual to program in extra thickness at the end of the shot for sag resistance.

8.1.6.3 Tooling Choices

With a basic understanding of the differences in the types of head tooling available, we can now look at calculations that will help make our tooling choices. It is important to keep these points in mind:

Parison diameter is dependent on head tool size, resin characteristics, regrind content, temperature, and punch-out speed.
Head tooling diameter is determined by part shape and size.
Head tooling diameter determines the size of machine required.

With emphasis on the second point, use the following formula to match the head tooling part.

$$\frac{2(W+T)}{3.14}$$

where $W =$ width of part,
$T =$ thickness of part.
This simple panel is 6 units wide by 2 units thick. Inserting these numbers into the tool formula will give us an approximate tool size required.
Example:

$$\frac{2(6+2)}{3.14} = 5.09 \text{ tool diameter}$$

From calculation a (5 in.) tool will be a starting point. The effect of blowup rate and extrusion rate on the total tool size is shown below. Blowup rate can be inserted into the formula D/B.

$$\frac{D}{B} = \text{Best tool for the job} \frac{5.09}{2}$$

where $D =$ tool diameter,
$B =$ blowup rate,
$C = 2.54 =$ proper tool for job.

Unfortunately not all parts are so simply designed as in this example. In cases where parts are more complex it is important to obtain an accurate measurement of the part periphery. Once that answer is found the original formula can be used. Use the following exercise to practice:

$$\frac{2(W+T)}{3.14}\frac{D}{B} \text{ Panel to be molded}$$

8.1.7 Part Weight and Wall Thickness Adjustment

Unlike many other molding processes, the part weight and wall thickness of an extrusion blow molded part can be adjusted up and down. Thus the performance of the part may be enhanced by putting the material where it is needed.

8.1 Extrusion Blow Molding

The method enables definition of performance goals and adjusting the process technique for maximum benefit.

8.1.7.1 Parison Programming

Moving mandrel programming is a technique that can shift wall thickness from heavier to lighter, weaker areas to fine tune the performance of the structure. With parison programming the relationship of die and mandrel is changed with hydraulic action to a precise, predetermined profile while the parison is being extruded (Fig. 8.16). This movement changes the gap between the die and mandrel, which changes the material thickness of the parison at various points. The parison is now ringed with sections of thicker or thinner material corresponding to shape of part.

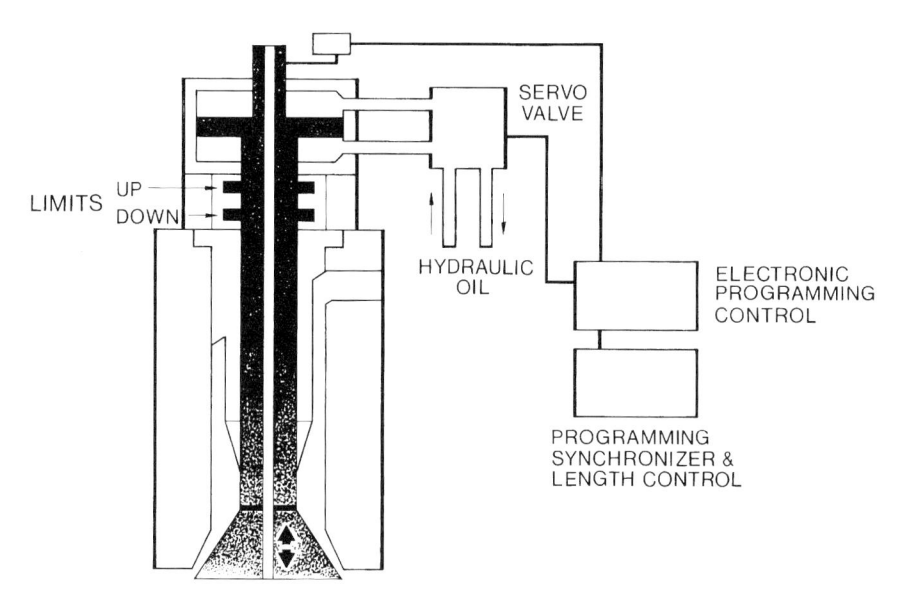

Figure 8.16 Moving mandrel programming

8.1.7.2 Die Ovalization

The parison material thickness also changes vertically (Fig. 8.17) as the material is extruded. Again sections respond to shape of the part. A further benefit of parison programming is reduced cycle time. This is governed by the thickest and thus the slowest cooling time for the part. By removing unneeded material from thick areas or reducing overall wall thickness, leaving thick areas only where needed, molding cycle time can be improved.

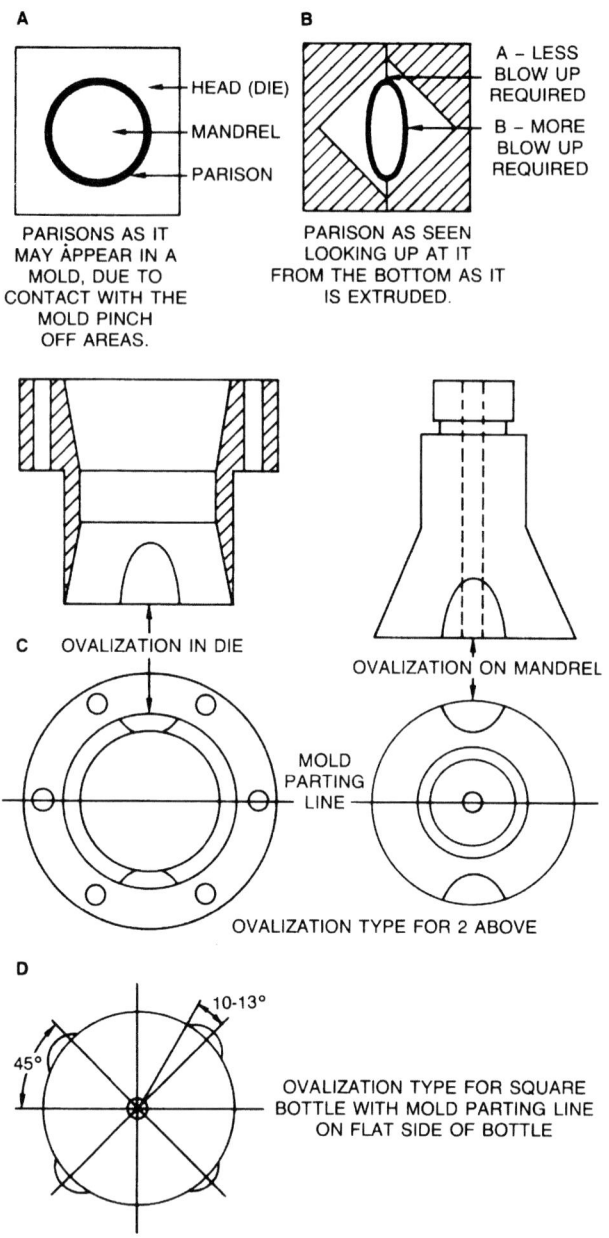

Figure 8.17 A schematic representation of ovalised tooling (Courtesy of Equistar Chemicals, LP.)

8.2 Blow Pins/Needles

The parison is inflated with air during the blow molding cycle via blow pins. This may be done through the head with the blow pin attached to the mandrel, which has already been indicated and is usual for bottles. For industrial and other products it may be more convenient or preferable to blow at the open end of the parison as the mold is closing (Fig. 8.18 a and b).

8.2.1 Needles

Blow needles are often used with parts that have complex shapes or several closed sections. Blow needles are smaller and can be inserted into the parison from the mold at almost any point or direction at the time of the mold closing (Fig. 8.18c).

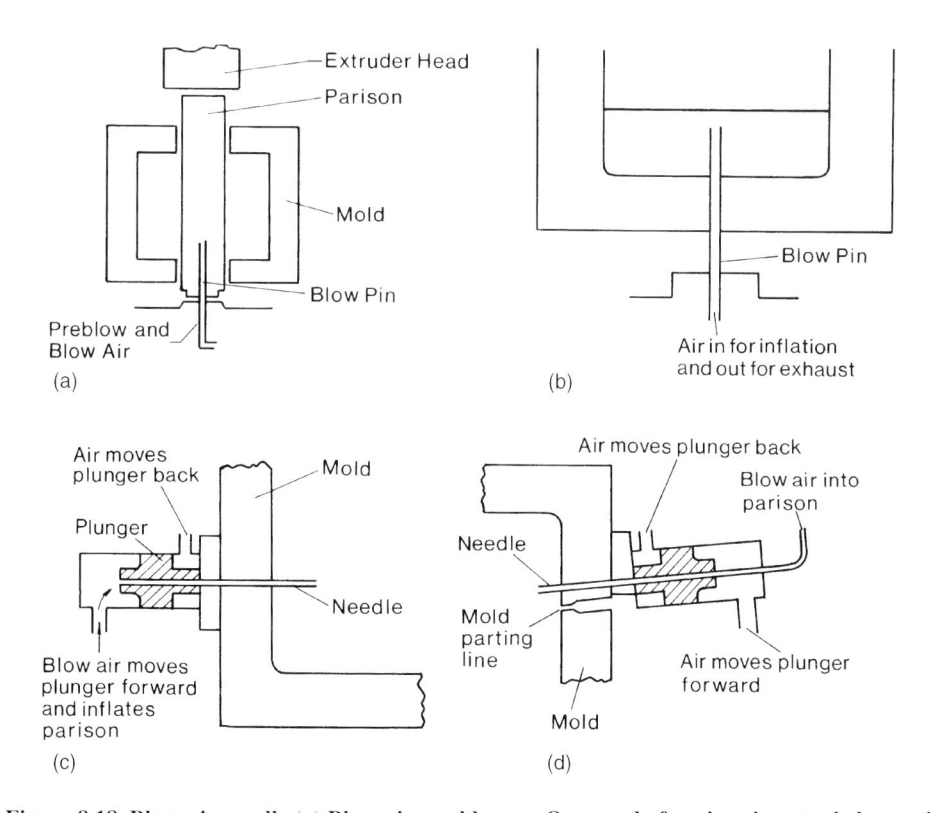

Figure 8.18 Blow-pin needle (a) Blow pin, mold open. Open end of parison is extruded over the blow pin. Molds close around blow pin and parison is inflated. (b) Blow-pin, mold closed. (c) Two-stage blow needle through the mold back. (d) Three-stage blow needle close to panel area

8.2.2 Pins/Needles

The size of the blow pin and the size and number of blow needles is important in both the speed of parison inflation and the speed of pressure reduction, part venting, or exhaust before the mold opens [3].

8.3 Injection Blow Molding

The name implies that the process uses the injection molding technique where the melted resin is forced under pressure through a manifold, sprue, and pin point gate into a mold cavity. A schematic cross-section is shown in Fig. 8.19. Though this is a ram type injection machine, it should be noted other types are normally used, such as a reciprocating screw.

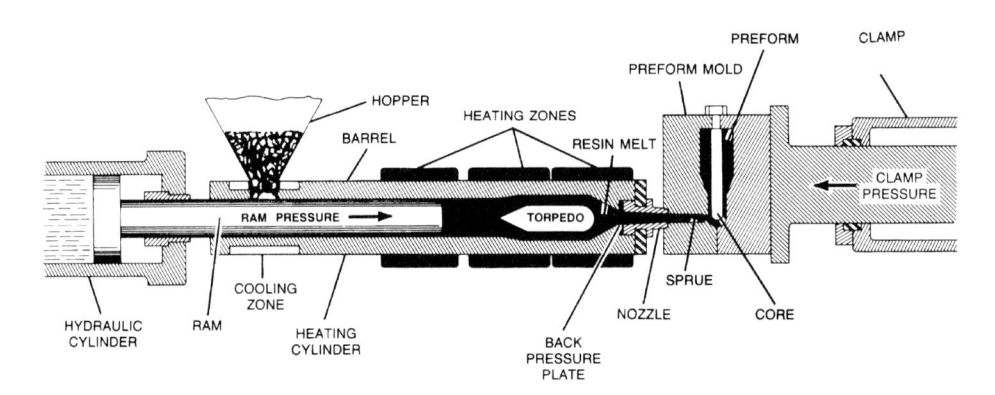

Figure 8.19 Cross section of a ram type injection molding machine for making preform (Courtesy of Equistar Chemicals, LP.)

8.3.1 Injection Blow Molding Process

This molding process comprises two steps (Fig. 8.20)

1. The injection molding of a parison on a support pin or core with the neck and threads already formed to their required dimensions, followed by
2. expansion of the hot parison, still on the support (core) pin, to its final shape.

Fig. 8.21 shows the four methods of injection blow molding. Farkus's method dates back to 1943, followed by the systems developed by Piotrowsky, Moslo, and Gusoni. The systems used today are basically developed from the Gusoni horizontal indexing method which was patented in Italy in 1961.

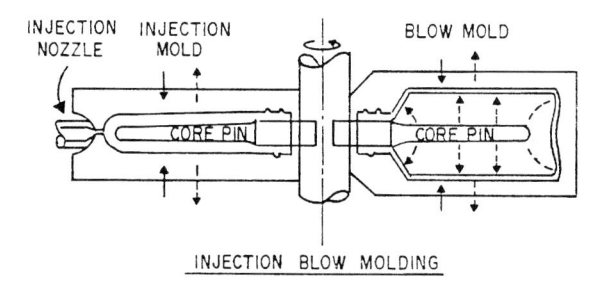

Figure 8.20 Injection blow molding

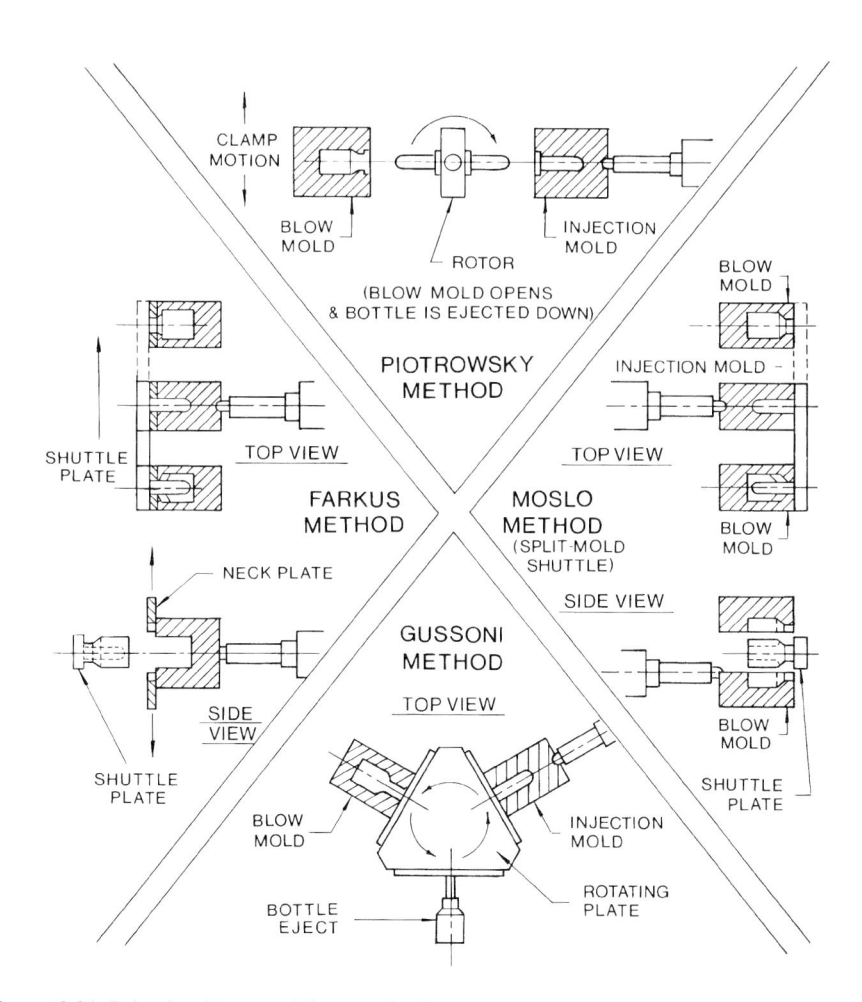

Figure 8.21 Injection blow molding methods

8.3.2 The Injection Blow Molding Machine

Injection blow molding is a technique mostly used to produce small containers, normally 373.2 grams (12 oz.) and under. A perform (parison) is injection molded onto a core rod in a split injection mold having the desired shape for blowing a finished part. While the injection molded preform is still in a semiviscous state, the mold is opened and the core rod containing the preform is transferred to a blowing mold. In the blowing mold, air is forced through the core rod and the preform is expanded to fit the cavity in the blowing mold. In contrast to extrusion blow molding, injection blow molding is always a noncontinuous, cyclical process (Fig. 8.19). Figure 8.19 also shows a cross-section of the machine with the preform mold [1].

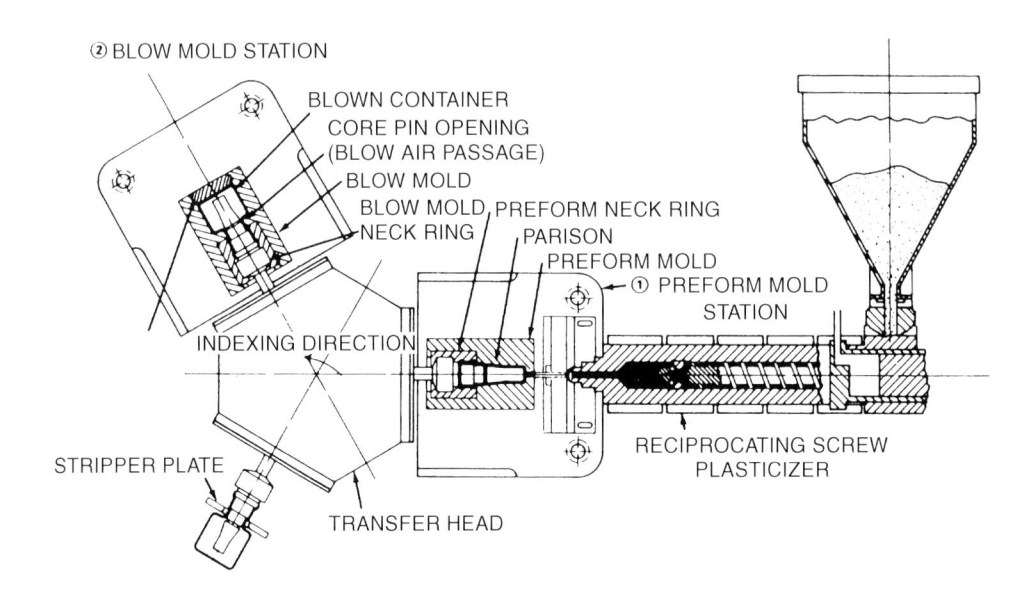

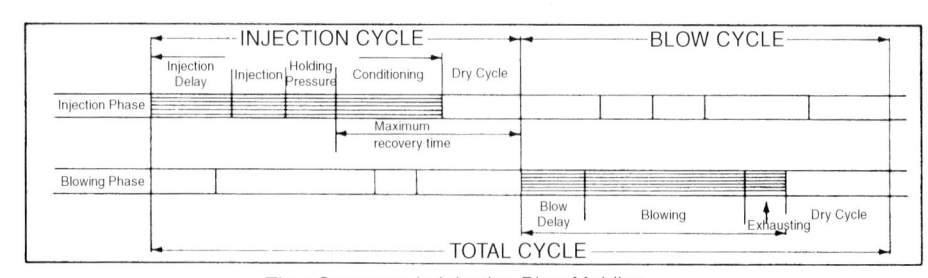

Figure 8.22 Process description of a typical three-station injection blow mold machine. (Courtesy of Equistar Chemicals, LP.)

8.3 Injection Blow Molding

Injection blow molding is usually a three-phase process with the first phase being the injection molding of a preform shaped as a test tube. The second phase is the expansion of the preform, with air pressure, into the bottle/form shape. The final phase is the removal of the part. All three phases take place at the same time [1]. A process description of a typical three-station injection blow molding machine is shown in Fig. 8.22.

The core rods are mounted on a turret mechanism turning, in sequence, from injection station to blowing station to ejection station. Economics dictate multicavity, matching injection and blow molds. Most machines are the three station types with 120 degree index turns between injection, blow, and ejection. Special four station machines are indexed on 90 degree turns with the fourth station used for special conditioning of the parison between molding and blowing (Fig. 8.23). Note that four station machines are used for two-color or multilayer bottles.

The primary advantages of injection blow molding are there is no scrap or flash resin to trim and reclaim, the neck finish and detail are of a very high quality, there is no process weight variation and, typically, it offers the lowest part cost for high-volume bottles of 373.2 grams (12 oz.) or less. It is important to note that a wide variety of resins can be processed; these include LDPE and HDPE (low and high density polyethylene), PP (polypropylene), PS (polystyrene), PVC [(poly)vinyl chloride], PET (polyethylene terephthalate), PAN (polyacryontrile), and others.

There are, of course, some limitations and disadvantages to the injection blow molding process in that the tooling costs are higher than for extrusion blow molding. Bottle sizes and shapes are limited to a container ovality ratio of 2:1 (HDPE), a blowup ratio of 3:1 (HDPE), and a diameter-to-height ratio of no more than 10:1. The slightly offset necks are

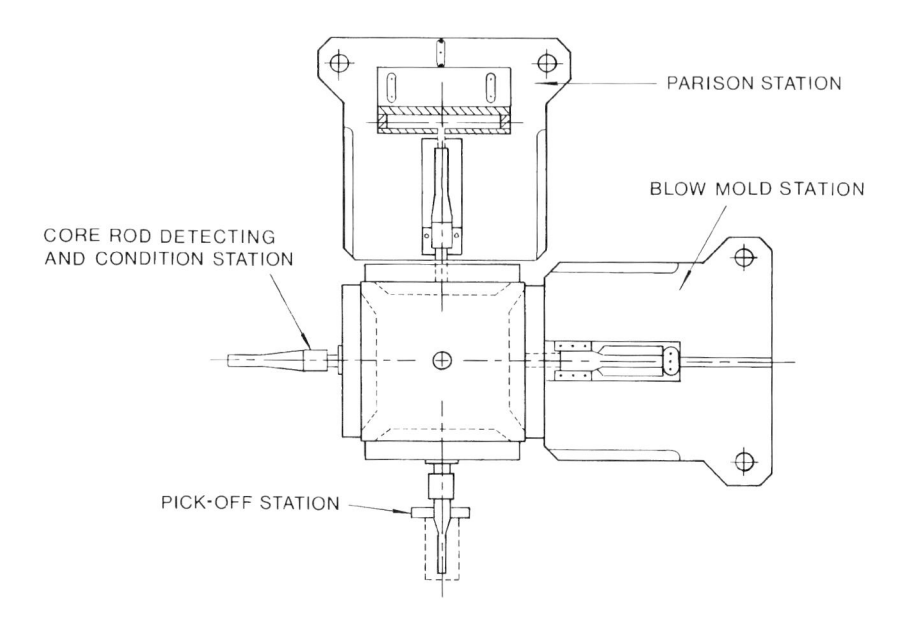

Figure 8.23 Four-station layout

possible with the injection blow molding process, but handles cannot be made. The process, however, when utilized appropriately, continues to provide a high-quality, cost-efficient end product [6].

8.4 Stretch Blow Molding

Stretch blow molding is basically a modification of either injection blow molding or extrusion blow molding; it involves the parison being conditioned to a specific temperature and then very quickly stretched in two (biaxial) directions. There are only four resins commonly used in stretched blow molding: PP (polypropylene), PVC [(poly)vinyl chloride], PET [(poly)ethylene terephthalate], PAN (polyacrylonitrile). While handlewear is generally not possible, and the machinery is highly specialized to the application and resin, stretch blow molding improves both resin and bottle performance; an illustration of stretch blow molding is shown in Fig. 8.24. A temperature-conditioned preform is inserted into the blow molded cavity, then rapidly stretched. Often a rod is used to stretch the preform in the axial direction with air pressure to stretch it in the radial direction. Some of the gains are impact strength, transparency, surface gloss, gas barrier, and stiffness.

8.4 Stretch Blow Molding 125

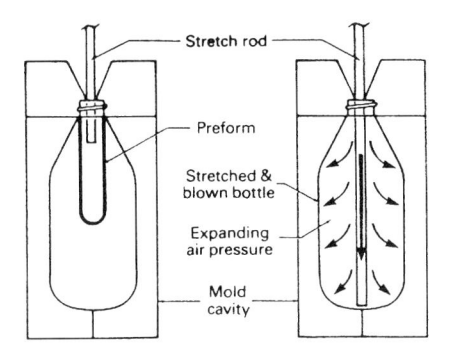

Figure 8.24a A temperature conditioned preform is inserted into the blow mold cavity, then is rapidly stretched. Often a rod is used to stretch the preform in the axial direction with air pressure to stretch the preform in the radial direction

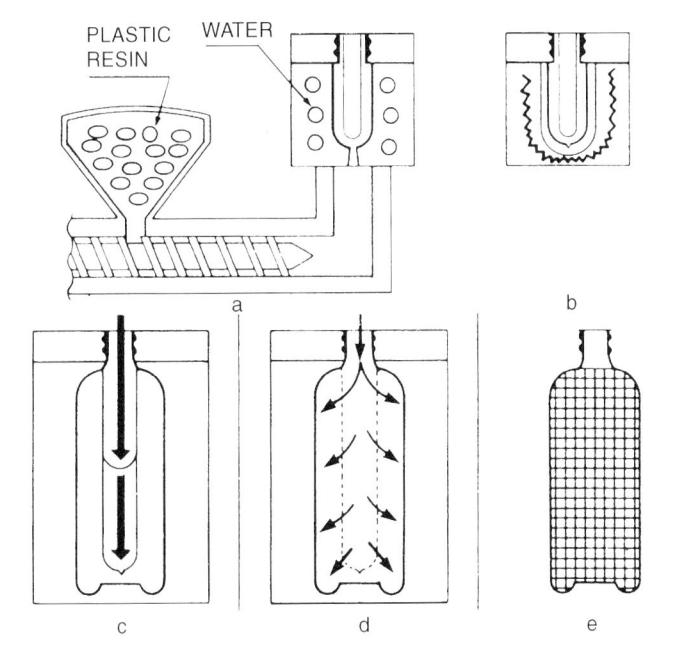

Figure 8.24b Single-stage PET process. (a) Milk-white material (crystallized PET chips) is melted by heating, then injected into the mold and rapidly cooled to form a transparent parison. (b) The cooled parison is reheated with an electric heater to soften it. (c) The softened parison is stretched to about twice its original length in the bottle mold. (d) Compressed air is blown into the stretched parison to expand it into the bottle mold. (e) The process has arranged the molecules of the materials in both lengthwise and crosswise orientations, making a finished product that is stronger and more attractive than those produced by older processes

8.5 References

1. Quantum U.S.I Division; Polyolefin Blowmolding and Operating Manual (now Equistar Chemicals, LP.)
2. *Packaging Technology—In House Blowmolding—An Opportunity For*, Irvin, C.
3. Fergunson, L. and Taylor, B. *Blow Molding, In Engineering Materials Handbook*, vol. 2, *Engineering Plastics*, Dostal, C.A. (Ed.) ASM International, Metals Park, OH
4. Belcher, S. Injection Blow Molding, In *Injection Blow Mold Handbook*, Lee, N. (Ed.) Van Nostrand Reinhold, New York

9

New Applications of Blow Molding Technology

This chapter deals with development of new processes in recent years so that the designer/engineer might consider new options and opportunities that are available. Three-dimensional molding; coextrusion blow molding of large parts; hard–soft–hard, soft and soft, hard and soft technology; long glass fiber reinforced blow molding; and blow molding foam technology are options that the designer/engineer could use.

9.1 Coextrusion Blow Molding of Large Parts

Coextrusion is used extensively in small bottles and containers. For large blow molded parts (18.92 liters or 5 gallons and larger) it is in its infancy. Its growth, however, will be very healthy and will accelerate in the coming years because of the need for recycling of plastic and the application of this technology to the manufacture of products requiring improved barrier properties. Some of these are discussed later in this chapter.

9.1.1 Reasons for Coextrusion

Large part coextrusion may be practiced for the following applications and benefits:

- Color savings (economic),
- decoration,
- view stripe addition in containers,
- surface treatment improved (for fluorination/sulfonation),
- surface improvements (engineering plastic/bonding agent/base material),
- recycling (2 or 3 layer structure), and
- barrier improvement.

Some examples of large part product applications for coextrusion are shown in Figs. 9.1 and 9.2.

128 New Applications of Blow Molding Technology

Figure 9.1 Large part co-extrusion parts (Courtesy of Krupp Kautex)

Figure 9.2 Large part co-extrusion parts (Courtesy of Krupp Kautex)

9.1.2 Typical Structures

Two-, three-, and six-layer structures are already in production in large part coextrusion. Environmental, economic, and recycling concerns and the accelerating need for improved barrier properties (e.g., in plastic gas tanks) will represent the largest growth areas in large part coextrusion in the near future. Figures 9.3 and 9.4 illustrate a variety of structures with their applications and benefits.

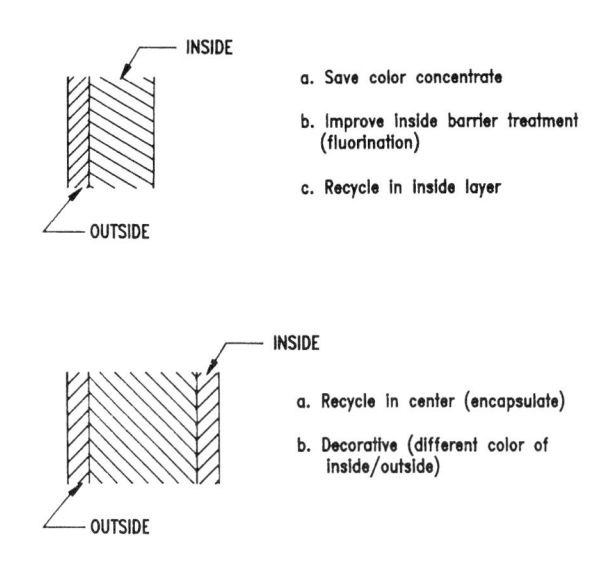

a. Save color concentrate

b. Improve inside barrier treatment (fluorination)

c. Recycle in inside layer

a. Recycle in center (encapsulate)

b. Decorative (different color of inside/outside)

Figure 9.3 Two-layer/three-layer cross-section (Courtesy of Krupp Kautex)

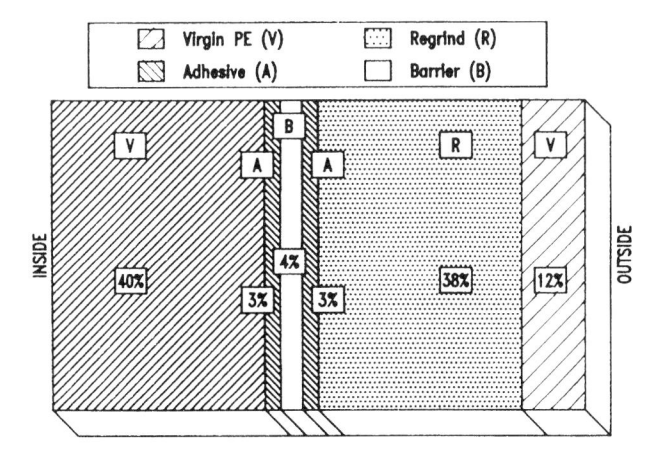

Figure 9.4 Typical fuel tank layer thickness in barrier co-extrusion (Courtesy of Krupp Kautex)

9.1.3 Intermittent and Continuous Extrusion Blow Molding

Two blow molding concepts are typically employed in the manufacture of large coextruded parts:

1. Intermittent blow molding and
2. continuous extrusion blow molding.

In intermittent blow molding an accumulator type head is typically employed whose accumulator may include a number of push-out pistons in accordance with the number of layers required. If the number of layers required is three or more, then the multipiston accumulator head may be used in conjunction with a number of reciprocating screws for formation of the extra layers (Fig. 9.5).

In continuous extrusion, a head similar to that shown in Fig. 9.6 is typically used. One obvious difference between the two concepts is that continuous coextrusion does not employ any reciprocating components for the formation of the various layers (pistons or reciprocating screws). The fact that it is continuous extrusion not only enhances the laminar flow of the various components of the structure through the head, but it also greatly simplifies the machine design and its controls. With continuous extrusion the wear and tear of moving components within the head is eliminated, color changing is greatly simplified, the complexity of the machine (controls) is greatly reduced (as is its operation), and the machine cost is reduced significantly.

For either concept, depending on the structure required and the materials used, each layer may be fed by its own extruder or the output of any extruder may be divided to form multiple layers of the structure. In the latter use, the versatility of the machine and its control is obviously reduced.

Table 9.1 compares the advantages of intermittent and continuous coextrusion. Clearly, continuous coextrusion is by far the preferred concept in terms of machine simplicity, stability and operation, energy efficiency, as well as most importantly machine and product cost. In addition, the cycle time of a continuously operating machine is less than that of an intermittent machine because the parison is already extruded and ready to be transferred into the blow mold. The parison must be ejected out of the accumulator head or

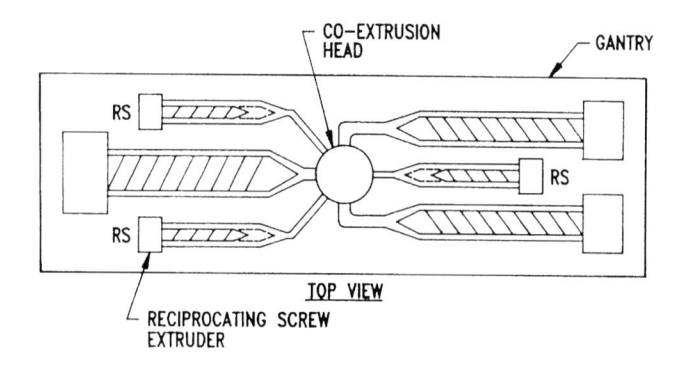

Figure 9.5 Location of extruders to co-extrusion head (Courtesy of Krupp Kautex)

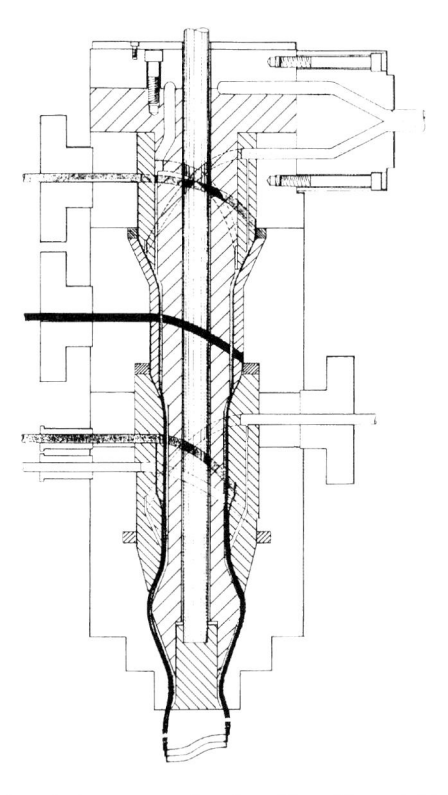

Figure 9.6 Krupp Kautex six-layer co-extrusion die with cadoidal manifolds (Courtesy of Krupp Kautex)

Table 9.1 Advantage Comparisons of Intermittent vs Continuous Co-Extrusion

	Intermittent	Continuous
Color changing		+
Machine heat-up time		+
Machine complexity		+
Machine cost		+
Time required to reach stability		+
Stability of operation		+
Cycle time		+
Minimum layer thickness capability		+
Operation requirements		+
Energy consumption		+
Range of material melt flows which can be processed	+	+
Product cost		+

(+) Advantage

reciprocating screw extruder slowly to preserve layer integrity. This is especially so in multilayer coextrusion of barrier structures in which the bonding and barrier layers may comprise only 1 to 4% of the structure.

9.1.4 Methods of Continuous Coextrusion Blow Molding

Continuous coextrusion may be practiced using one of the following two methods:
1. Shuttling clamp or
2. parison transfer system.

Of the two concepts, the parison transfer system is much preferred since it requires less energy to transfer a 45.36 to 136.08 kg (10 to 30 lb) parison vertically down into the mold than it takes to shuttle a heavy clamp and mold. In addition the parison transfer concept runs on a faster cycle time than the shuttling clamp concept.

Figures 9.7 and 9.8 illustrate the parison transfer continuous coextrusion concept. A three-layer continuous extrusion blow molding machine is shown in Fig. 9.9. Note that the head must be raised 0.9144 to 1.524 m (3 to 5 feet) (depending on product) above the top of the mold to allow for the continuous extrusion of the parison. At the start of the cycle, the multilayer parison is continuously extruded. When it is the correct length (the extrusion speed is timed with the cooling cycle time of the particular part), the parison gripper clamps the parison directly underneath the head and transfers it down into the mold where it is blown. Simultaneously, the next parison is continuously extruded in preparation for the

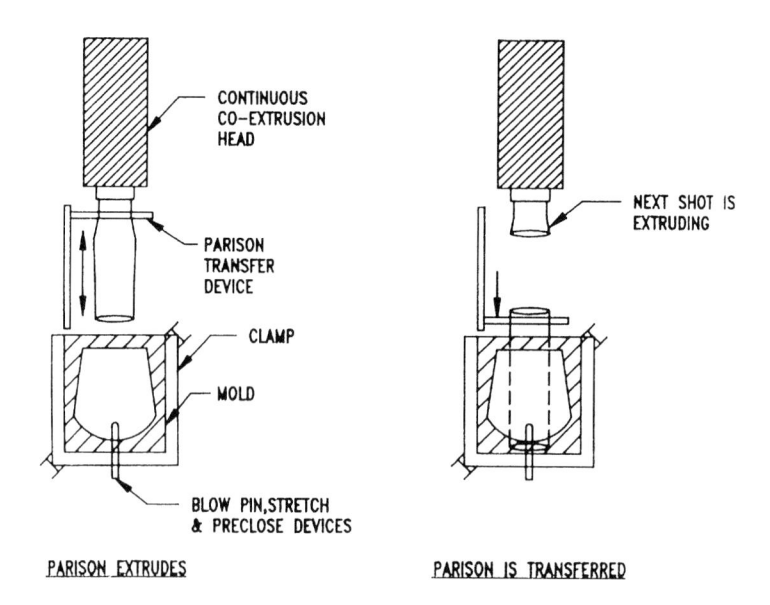

Figure 9.7 Continuous co-extrusion with parison transfer device

9.1 Coextrusion Blow Molding of Large Parts   133

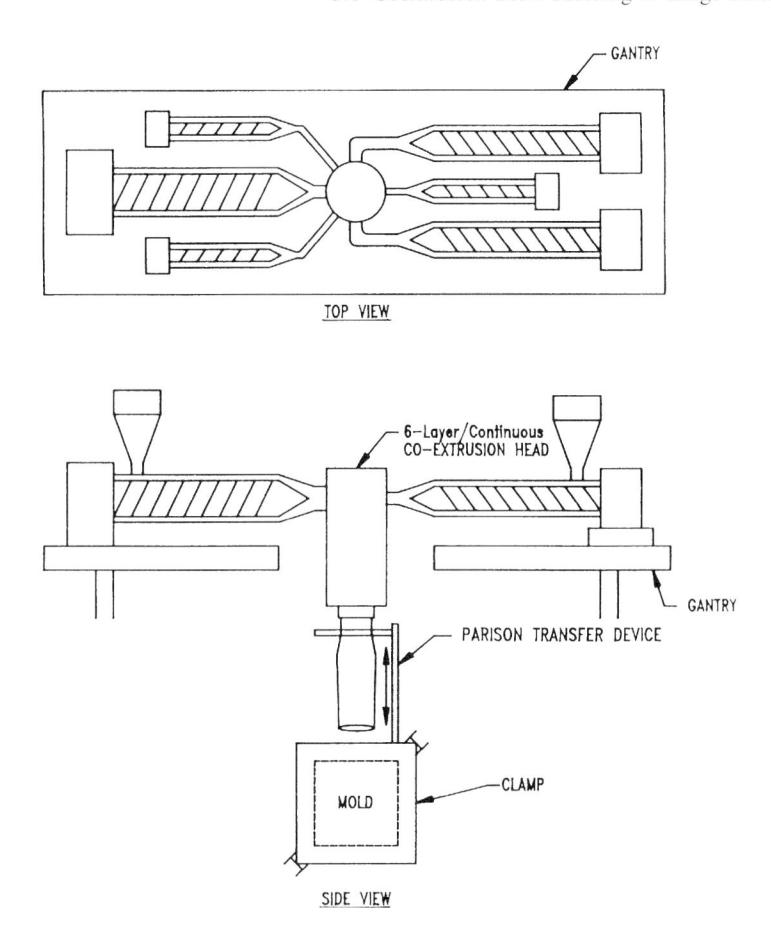

Figure 9.8 Six-layer/six extruder continuous co-extrusion machine

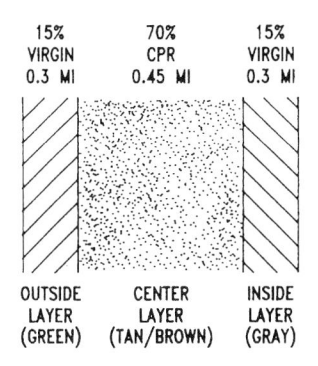

Figure 9.9 Three-layer HDPE structure, 32 gallon trash can

next cycle. Once the mold is closed, the gripper transfers vertically to clamp the next parison and repeat the cycle. Figure 9.9 shows a section through a can wall.

Continuous coextrusion machines are installed and capable of manufacturing multilayer parts up to 227.1 liters (60 gallons) in size. Parison hang times of as long as $2\frac{1}{2}$ minutes are employed (depending on cooling cycle) in the manufacture of multilayer gas tanks when using polyethylene having melt flow in the range of 5 to 10 ultra high molecular weight material (HLMI). Multilayer product shown in Figs. 9.10 and 9.11.

Figure 9.10 Multilayer trashcan product (Courtesy of Krupp Kautex)

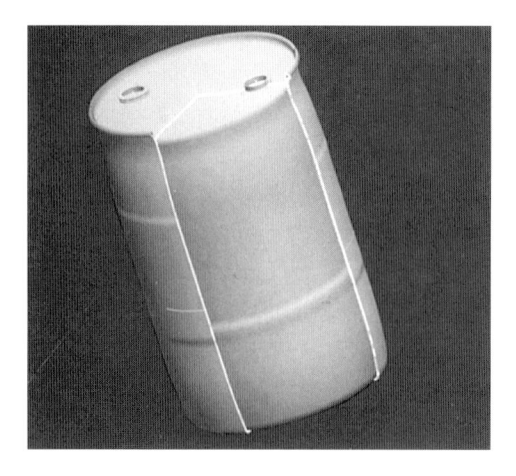

Figure 9.11 Section through multilayer drum

9.2 Three-Dimensional Blow Molding

The three-dimensional blow molding concept was developed several years ago in Japan. The most successful of these technologies is the Placo X–Y machine which moves the mold under the head [2]. There are many advantages to three-dimensional blow molding including minimal flash, seamless parts, and sequential extrusion. Many complex shapes can be easily produced using the three-dimensional blow molding process (Fig. 9.12). At this time three major machine manufacturers offer process options for three-dimensional molding: X–Y processes, suction blow molding, and curved blow molding.

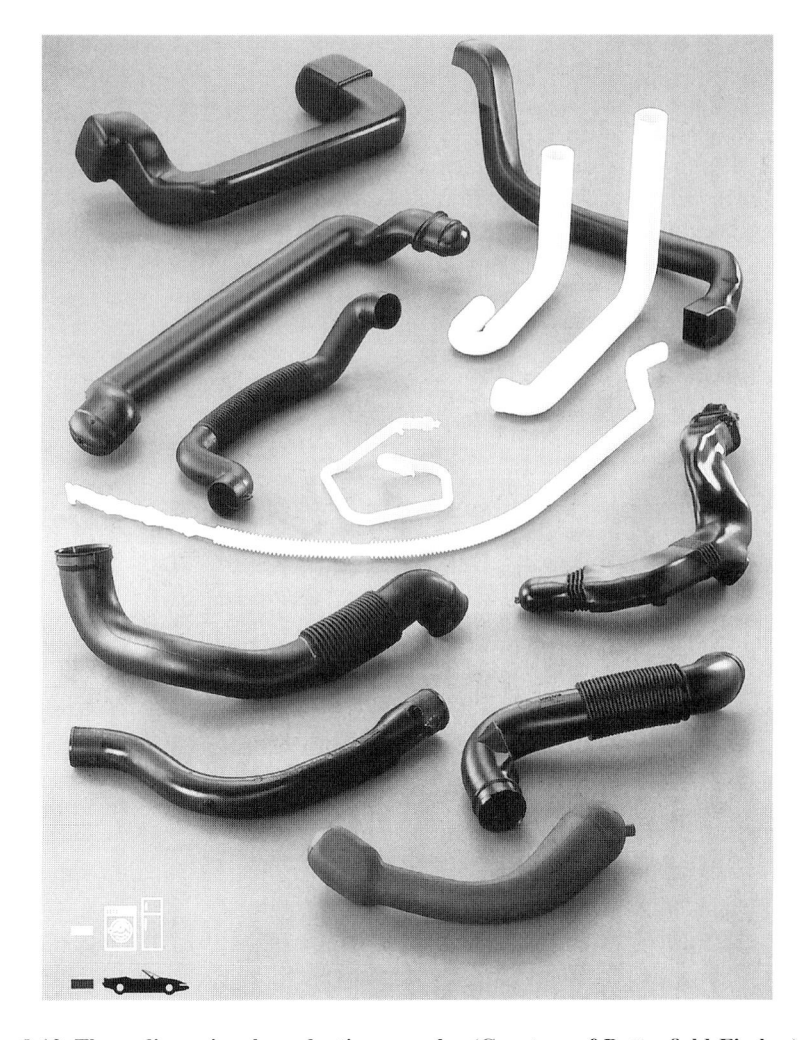

Figure 9.12 Three-dimensional production samples (Courtesy of Battenfield Fischer)

9.2.1 Mold Inclining System and Computer Controlled Mold Oscillating Device

The Placo X–Y machine utilizes a platen system that is either fixed at a 45 degree angle or adjustable from 45 to 90 degrees. The bottom or right-hand platen moves in the X and Y planes to position the mold under the head to allow the parison to be laid directly into the cavity. The parison is prepinched at the head and the mold moves under the head as the parison extrudes. This allows most of the plastic to be contained in the mold and flash produced only on the very ends of the part. In automotive duct applications, this area is scrap anyway because the blow domes are trimmed away to allow both ends of the part to be open.

9.2.1.1 The X–Y Process

The X–Y process is as follows: The bottom or right platen moves in the X and Y planes so that the parison follows the shape of the cavity. The bottom platen registers home and the top or left platen closes onto the bottom plate. The blow needle then fires and a normal blow molding cycle begins. The top platen opens and the bottom platen shuttles in the Y plane toward the operator then the part may be removed by a person or by robot. The bottom platen travels to its highest point and the cycle starts again.

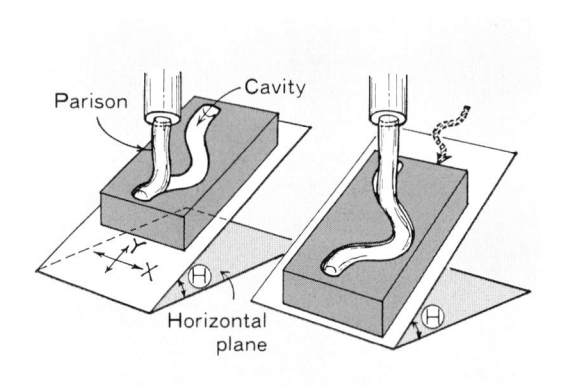

Figure 9.13 Computer controlled operating device: The mold can be oscillated freely in two-dimensional direction within the inclined mold platens (Courtesy of Placo, Ltd.)

9.2.1.2 Formed Parts

Parts formed with the X–Y process have consisted (primarily) of duct work for automotive applications. These ducts are excellent candidates for the X–Y process because they are normally hollow tubes with several bends. A conventional blow molding machine would produce two to three times the part weight in flash while producing these parts. Since the amount of flash is reduced, the amount of secondary work is also reduced, with savings in auxiliary trimming equipment being realized. Two or three materials may be used with the process depending the number of extruders on the machine.

9.2.1.3 Features of the X–Y Machine

A feature of the X–Y process is the ability to produce sequentially extruded parts. They are produced by using one extruder for a segment of the part, then using another extruder for a second segment. Parts can be made with thermoplastic elastomer and a polyolefin in combination to allow the ends of the parts to be flexible, while rigid areas for strength and flexible sections in the middle can be arranged to allow for vibration and ease of assembly. Many separate parts can thus be combined into one for added value, eliminating many hose clamps and connections in the process.

Another feature of X–Y blow molding is that very even wall thickness may be achieved. Because the parison is continuous within the cavity, there is no pinch-off seam.

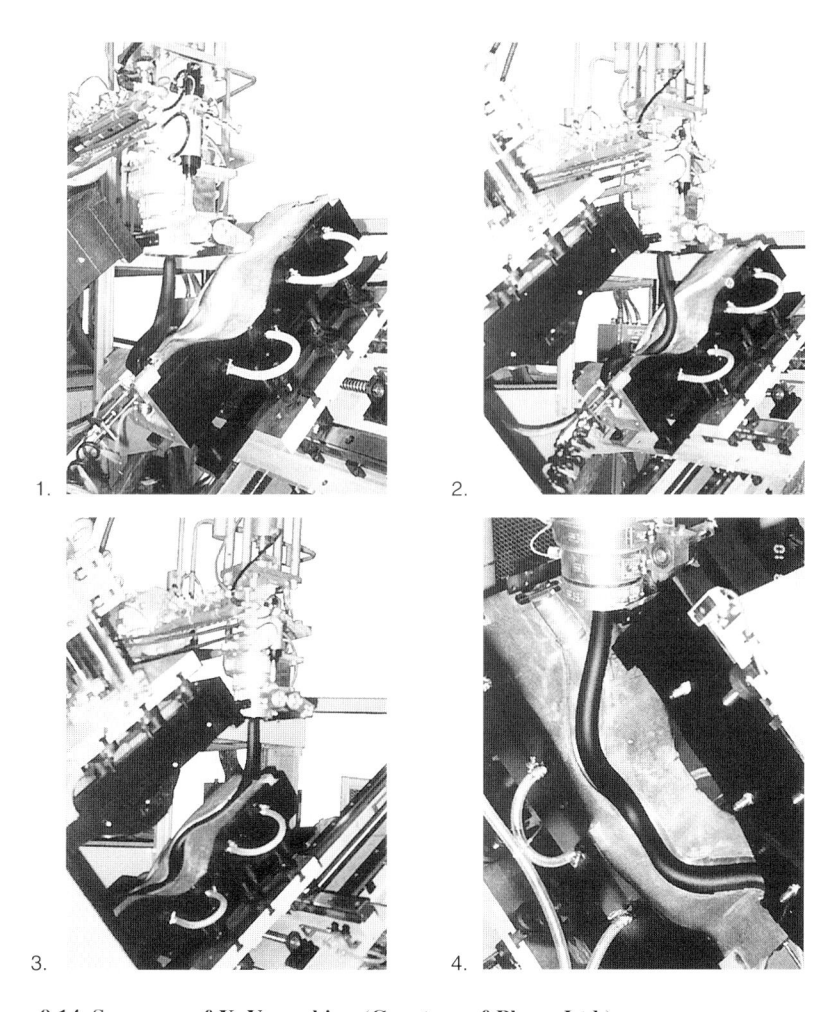

Figure 9.14 Sequence of X–Y machine (Courtesy of Placo, Ltd.)

This reduces molded-in stress and increases physical integrity in situations involving pressure. The parts also have better aesthetics, with minimal scarring from trimming.

If brackets or outlets are required on the X–Y parts, injection molded inserts may be molded in during the cycle. Because most parts run with the platen in the tilted position, the inserts easily stay in place during the molding cycle. Also, because the bottom platen moves toward the operator, insertion of these parts in the mold is simplified. Compression molded flanges may also be formed on the X–Y parts by programming the cavity to move so that the parison is partially captured between the mold halves to form the flange (Fig. 9.14).

Other current applications include agricultural ducts, automotive filler pipes, and furniture armrests. Armrests incorporate another variation of the X–Y process. A complex shape can be formed with a rigid inner layer and a flexible outer layer for the feel and texture. Also, a deep drawn double-walled article may be blown (Fig. 9.15).

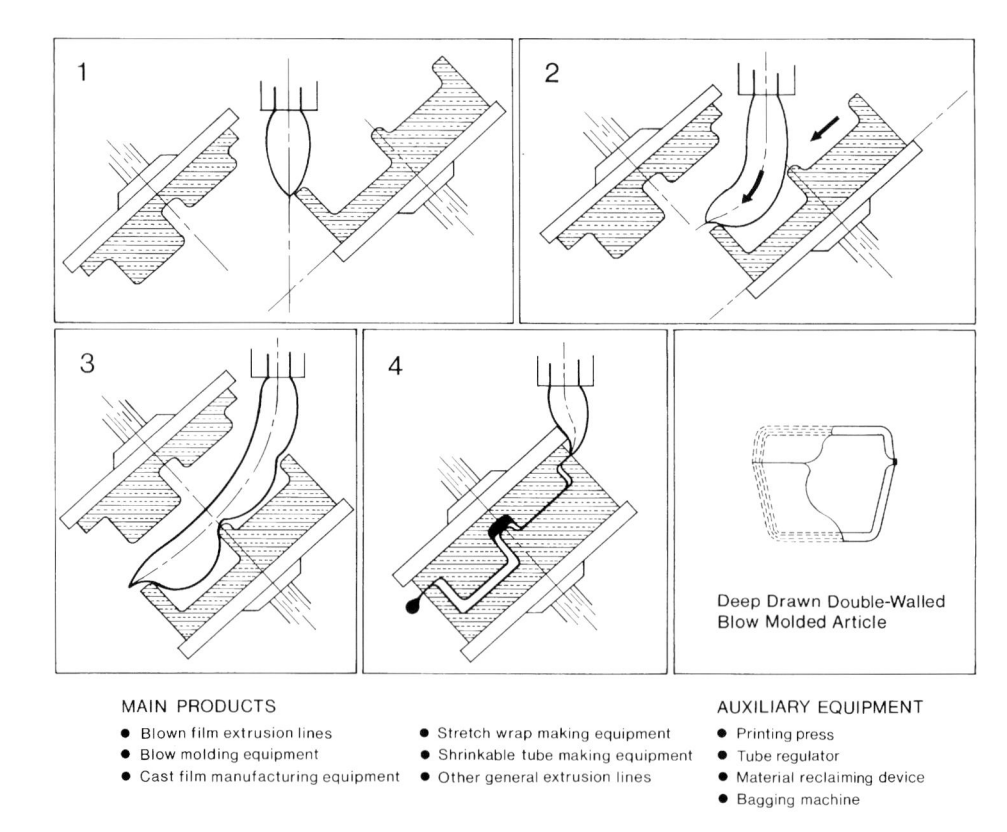

Figure 9.15 Deep drawn double wall article (Courtesy of Placo, Ltd.)

9.2.2 The Three-Dimensional Technology of Suction Blow Molding

Battenfield Fischer has extended its three-dimensional technology range by adding the suction blowing molding process [3]. Using this process, the parison is sucked into the closed blow mold, requiring ejection of the parison and extraction of the air volume.

The blow mold consists of a main section as well as upper and lower sliding core segment (Fig. 9.16). The process has the following steps:

1. The blow mold closes.
2. The suction device, which is part of the bottom segment, is ready for operation once the mold is closed.

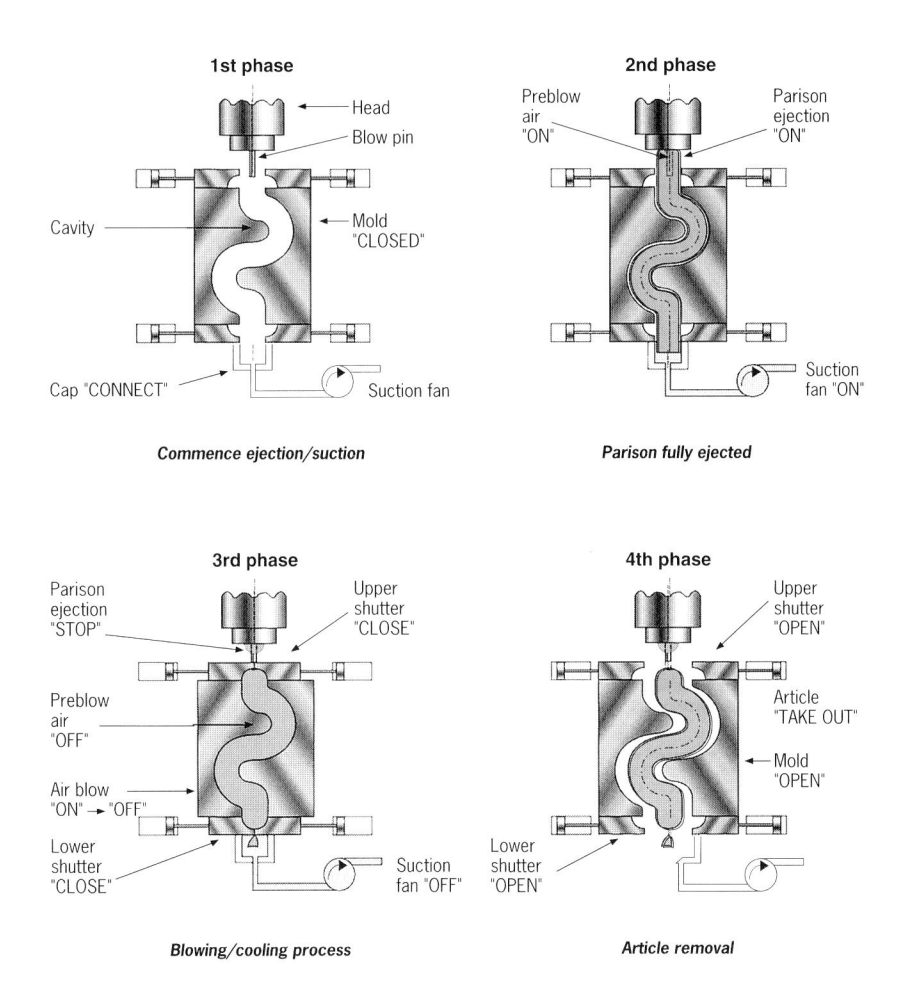

Figure 9.16 Basic operation diagram (Courtesy of Battenfield Fischer)

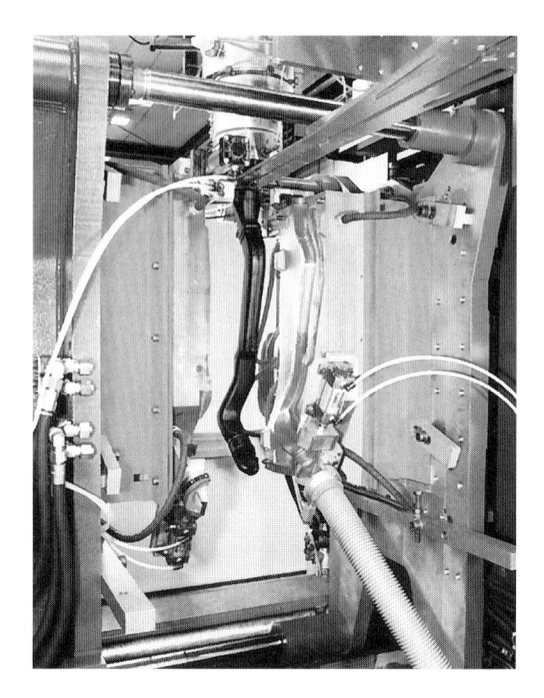

Figure 9.17 Suction blow molding: article discharge (Courtesy of Battenfield Fischer)

3. The parison ejection process begins simultaneously with the suction function.
4. The parison is preblown with support air.
5. As the ejected parison reaches the desired length, the ejection and suction processes are stopped automatically. At the same time, the upper and lower sliding segments close.
6. The article is blown by a blowpin or needle.

At the end of the cooling time, the mold opens and the article is removed (Fig. 9.17).

The suction blow molding process enables low-scrap production of three-dimensional bent articles with the following advantages over other process:

There is lower investment in machines and molds; therefore ideal for shorter production runs [Just-In-Time (JIT)].
The technology is simple and easy to handle.
As the mold is closed during the cycle unwanted squeezing can occur.
As the airstream in the mold prevents premature contact between the parison and mold surface, the finished article has an even surface.
Even materials with lower melt strength, for example polyamides, can be processed.
The process can be adapted to sequential coextrusion (SeCo) as well.

The BFB machine series BFN2 and BFB8 are used for this process, and full details may be obtained from the manufacturer.

The BFB-3D process with parison manipulation has significant advantages when producing complicated shapes in articles demanding standards concerning wall thickness distribution and reproducibility, but requires additional investment in comparison to the basic suction blow process.

9.2.3 Three-Dimensionally Curved Blow Moldings

Substantial savings are possible with three-dimensionally curved parts. They may be produced by using blow molding machines without a weld seam specifically designed for the production.

In this process the extruded parison is moved directly into the mold cavity so that the remaining welding seam length is reduced to a minimum [1].

The clearance in the mold mounting area of Krupp Kautex machines (Fig. 9.18) permits mounting complex blow molds and operating them with several individual mold part movements. Programming manipulates for a three-dimensional handling of the parison and customized devices permits the production of complex parts without welding seams.

Krupp Kautex has installed a machine for the commercial production of EPDM (ethylene–propylene–diene monomer) air ducts with venting sockets. While EPDM has several excellent properties for car manufacturers it was considered difficult to process and virtually impossible to weld. As a consequence, attempts to weld inserts such as pipe connection pieces in EPDM processing failed to produce satisfactory results.

Figure 9.18 Kautex three-dimensional machine (Courtesy of Krupp Kautex)

9.3 Hard–Soft–Hard and Soft–Hard–Soft Technology [1]

It has been shown that a combination of various materials imparts specific properties to a part in extrusion blow molding. In this process the extruded parison is built up in a radial direction by several overlaying layers (traditional or radial coextrusion).

New in practical application, however, is the use of various materials in the axial direction, that is, specific article sections may be provided with specific properties by choosing corresponding materials.

9.3.1 Axial Coextrusion

In axial coextrusion, a single article can be made up of sections with different hardness, elasticity, strength, or deflection temperature under load. This option opens up a multitude of new applications such as air suction pipes or air ducts. For such products often contradictory properties are required in one and the same article. So the middle section may call for stability against excess pressure and vacuum while the pipe ends need to be flexible to permit tight sealing under tube clamps as well as easy fitting (Fig. 9.19).

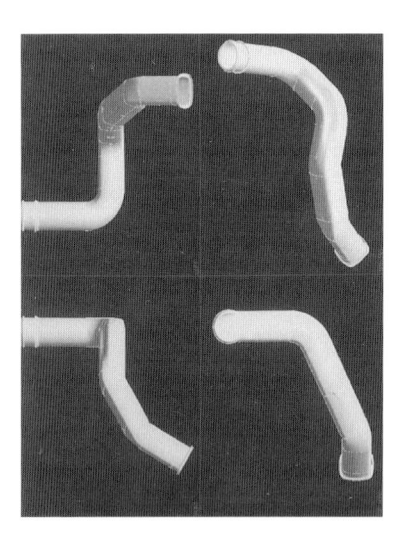

Figure 9.19 Hard–soft–hard examples (Courtesy of Krupp Kautex)

9.3.2 Preferred Material Combinations

Preferred material combinations for this hard–soft–hard or soft–hard–soft coextrusion technology are PP and EPDM or PA with elastomer modified PA, depending on the ambient temperature.

In addition to the above mentioned intake and air ducts, typical applications also include protective covers, clean air tubes, shock absorbers, hot air tubes, and various connection pipes not only for the automotive industry but also for the household appliances industry.

If axial coextrusion is combined with production without a welding seam, the resulting blow molding technology is particularly suited for the production of parts that until recently had to be produced from rubber materials. The substitution of rubber parts by TPE blow molding has many advantages.

In addition to shorter cycles and lower energy consumption, the improved properties of TPE parts permit a reduction in the wall thickness, which yields an immediate cost advantage given comparable raw material costs. The simple recycling of TPE is another bonus as no disposal problems arise.

9.4 Long-Glass-Fiber-Reinforced Blow Molding

The use of short-glass-fiber-reinforced thermoplastics (mainly PA 6, PA 6.6, and PP) with medium fiber lengths between 0.2 and 0.4 mm (0.0078 in. and 0.0156 in.) has been known for some time. Unfortunately it was impossible to realize improvements in tensile strength, modulus of elasticity, and deflection temperature under load in the order hoped for by using glass fibers. The reason was the insufficient length of the fibers, which was reduced substantially from its original length as a consequence of the shearing action in the extruder.

9.4.1 Breakthrough

Using a special process technology it is possible to achieve average fiber lengths of about 10 mm (0.3937 in.). This difference from the length of short glass fibers of 0.4 mm (0.0156 in.) in average results in significantly improved properties [1].

9.4.2 15% Long-Glass Fiber

A 15% wt long-glass-fiber additive yielded almost double the tensile strength and modulus of elasticity of conventional short-glass-fiber-reinforced PP. With a deflection temperature under load of 130 °C to 140 °C (266 °F to 284 °F), long-glass-fiber-reinforced PP is particularly suitable for applications in the motor compartment. Several interesting applications are also being evaluated for interior applications because long-glass-fiber reinforced components offer a low coefficient of thermal expansion and good sound absorbion that make such materials interesting for dashboards and inner panels.

9.5 Blow Molding Foam Technology [1]

Krupp Kautex, Borealis, and OBG Design has introduced foam technology, based on a patent held by leading equipment manufacturing Krupp Kautex (Patent Number US 4.874.649) (Fig. 9.20).

A special-recipe Mastermix from Borealis, a major polyolefin producer, allows highly reliable processing and consistent results. Customized product designs are available from OBG Design. Examples of transportation pallets are Figs. 9.21a and b.

Figure 9.20 A Krupp Kautex blow molding machine incorporating blow molding foam technology add-ons processing a pallet based on a design from OBG Design.

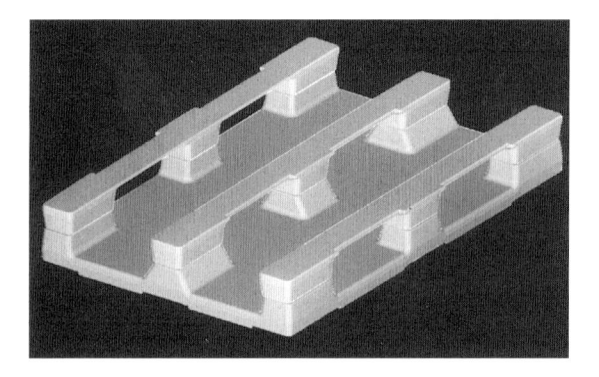

Figure 9.21a (Courtesy of Krupp Kautex)

Figure 9.21b (Courtesy of Krupp Kautex)

9.5.1 Advantages

Blow molding foam technology (BFT) offers a number of very specific advantages to the converter (Fig. 9.22). These include:

- Increased stiffness,
- a choice of surface design and geometry through the use of polymer top layers,
- savings in handling weight at given mechanical characteristics,
- the use of standard extrusion blow molding equipment, with add-ons for BFT process, and
- access to a new range of polyolefins for different foam applications (HDPE, LLDPE, LDPE, and PP).

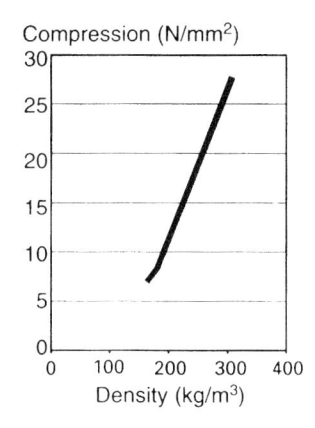

Figure 9.22 Effect of the foam on compression (Courtesy of Krupp Kautex)

146 New Applications of Blow Molding Technology

Step 1. Pallet*

Step 2. Pallet with containers

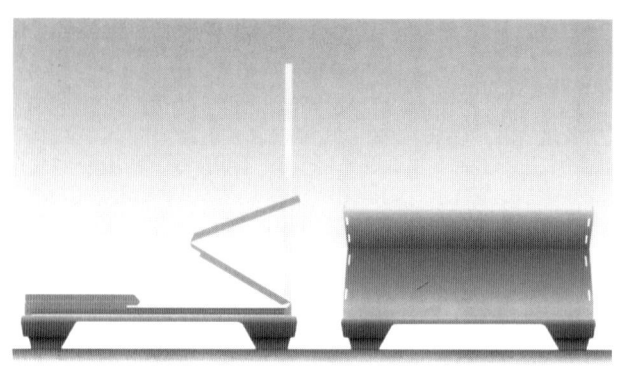

Step 3. Pallet in a collapsing stage

* Pallet for light & heavy weight articles between 9 kg and up. The requirements specification and design gives the characteristic of the pallet weight

Step 4. Pallet with collapsed container

Figure 9.23 BFT collapsible pallet container system (Courtesy of Krupp Kautex)

9.5.2 Blow Foam Technology Products (Fig. 9.23)

BFT offers a variety of products, such as pallets, collapsible side-wall containers, furniture, panels, automotive parts, and collapsible boxes. BFT offers substitution potentials such as:

EPS (boxes/containers)
Carton (boxes/containers)
Wood (pallets/boxes)
Steel (containers)
Plastic (pallets/boxes)

9.6 Conclusion

The many options such as three-dimensional blow molding, coextrusion blow molding of large parts, and long glass fiber-reinforced blow molding, to name a few, has proven very beneficial for the designer/engineer. These new options have brought cost savings, surface improvements, and many other benefits to blow molding.

9.7 References

1. Antonopoullas, J. Krupp Kautex Maschinenbau GmbH Kautexstrasse 54, Bonn Germany, and Edison, NJ
2. Platco, Co. Ltd., Iwatsuki Satama Japan, Hobson Bros., Torrance, CA, Shell Rock, IA
3. Battenfield Fischer Blow Molding Technology, Battenfield Fischer Blasformtechnik GmbH Postfach 11 63, 53821, Transdorf Spich, Bonn Germany

10
Understanding the Mold

The mold determines the shape of the end product with all its details. It helps to provide the end product with essential physical properties and the desired appearance. Usually, the mold maker builds the blowing mold according to the molder's or his customer's specifications. But frequently, minor adjustments or improvements that would not justify its being returned to the mold maker can be made with equipment and knowledge available in the blow molding shop.

10.1 Main Characteristic of Mold Halves

The blowing mold may have a number of parts, counting its various inserts, but it usually consists of two halves. When closed, these halves will form one or more cavities that will enclose one or more parisons for blowing. For bottles and containers the two mold halves are alike. Industrial parts can be complex, however, including slides and inserts. There are usually no male and female sections, an exception being double-walled case molds.

Pinch-off edges are generally provided at both ends of the mold halves for bottles. A blowing pin may have the additional function of shaping and finishing the neck inside. Both mold halves must have built-in channels for the cooling water. Sets of guide pins and bushings or side plates in both mold halves ensure perfect cavity alignment and mold closing. Accurate guiding devices in both mold halves reduce setup time. Figure 10.1 shows the two halves of a blowing mold for small bottles. Figure 10.2 shows the location of the cooling water channels.

On some blowing presses, mold closing is carried out in two steps, first at high speed, with lower pressure to approximately, 6.2 to 13 mm (1/4 to 2 in.) "daylight" and then slower, with higher pressure, to protect the mold from tools and/or to strengthen pinch-off weld (Fig. 10.9).

Molds are not necessarily positioned vertically, that is, in line with the parison. They may occasionally be tilted. This will result in a nonuniform distribution of resin which may be helpful, for instance, when such irregular pieces as a pitcher with a handle are being blown. It may also result in some saving in parison length.

150 Understanding the Mold

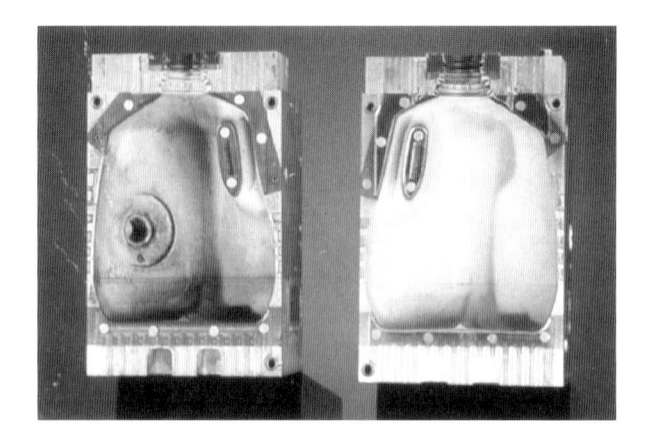

Figure 10.1 Blow mold halves (Courtesy of Johnson Controls, Manchester, MI)

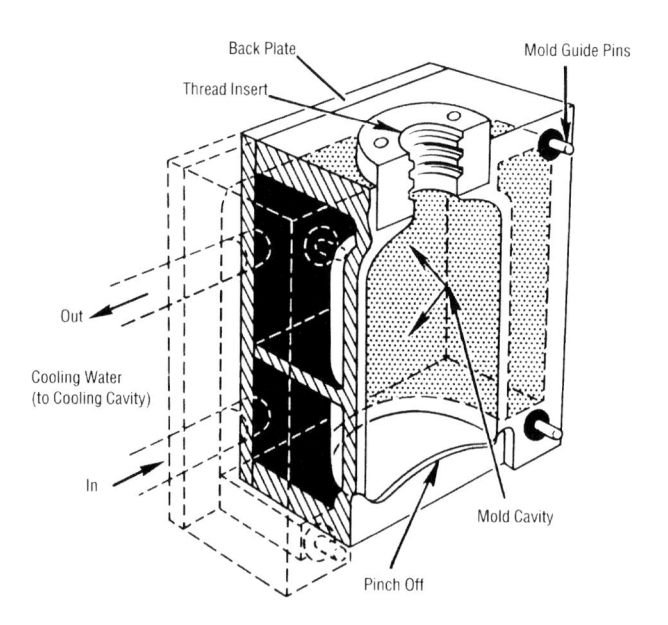

Figure 10.2 Schematic of a blow mold half, with the cooling-water channels indicated

10.2 Mold Materials

Because of the comparatively low clamping and blowing pressure, the blowing mold need not be made of high-tensile strength material, with the possible exception of molds for very long production runs, say, hundreds of thousands or millions, which are sometimes made of steel. The predominate raw materials for blowing molds are beryllium copper, cast aluminum alloys, zinc alloys such as Kirksite, and aircraft grade aluminum thick plate. All these alloys are excellent materials for blowing molds.

10.2.1 Cast Aluminum and Beryllium

Cast aluminum and beryllium–copper molds may be slightly porous, and, occasionally, blow molders have experienced some permeability of such molds to the viscous resin. This may affect the appearance of the blown part. The remedy is casting stainless steel tubes for water lines on the inside back face of castings.

10.2.2 Aluminum Plate

Aluminum aircraft plate or blocks are now commonly used because they are as hard as mild steel and contoured cavities can be readily machined with computer aided machining.

10.2.3 Steel

Steel molds are heavier, more expensive, and more difficult to machine than those made of nonferrous alloys. Higher weight will mean more setup time in the molding shop. Moreover, the heat conductivity of steel is inferior to that of the three nonferrous mold materials. This results in a slower cooling rate and a correspondingly longer cooling cycle and consequently, a lower production rate for steel molds.

10.3 Importance of Fast Mold Cooling

Fast heat transfer of the material of which the mold is made is of utmost importance because, as explained, the cooling step controls the length of the blow molding cycle. (Cooling takes up roughly two thirds of the entire blowing cycle.) Good heat transfer means faster cooling, and faster cooling means more items blown per hour, that is, less expensive production. This is the main reason why, for blowing molds, the above mentioned alloys are generally preferred to usually more durable steel.

10.3.1 Heat Transfer Rate

Considering only their heat transfer rate, the principal blowing mold materials follow each other in this order:

Thermal Conductivity (BTU in/ft^2 H°F)
Beryllium–copper	CA172	770
Kirksite	A	640
Aluminum	7075 T6	900
Steel	AISI P.20	200
AISI 420 Stainless		166

Occasionally, several different alloys are used in the same mold to obtain desired strength and special cooling conditions. However, as these mold materials have different heat transfer rates, a blowing mold, with the exception of the steel pinch-off inserts (see pinch-off section) should be made of only one material. Different materials with consequently different heat conductivities at various points of the mold will result in nonuniform cooling. This, in turn, might set up areas of stress in the finished piece that are susceptible to splitting in use.

10.3.2 Cooling

The blowing mold halves must be adequately cooled to solidify the part quickly, immediately after the parison has been blown out against the mold walls. The cooling should be chilled by a heat exchanger to 4 to 20 °C (40 to 70 °F). Such low temperatures may, however, cause water condensation on the outside mold walls. Usually, the cooling water is recirculated, that is, reused time and again for a long period. Sometimes, it is partly recirculated and mixed with fresh tap water, to maintain the desired temperature and to economize.

10.3.3 Cooling Lines

In cast molds the water usually circulates through the hollow mold halves. Sometimes, as previously mentioned, the tubing system is cast into mold. However, to create the most useful flow, water channels are machined and/or drilled into the mold halves. Well placed channels will ensure that the cooling water comes as close to the mold cavity as feasible. Cooling channels should also be as close (in length or other measurements) to the parting lines caused by the separation lines of the two halves or by inserts. Parting lines will practically always show along mold separation lines. Cooling these areas will result in better finish of the piece along the parting lines. The water lines are usually 11.1 to 14.3 mm (7/16 to 9/16 in.) in diameter and should have several right angle turns to help turbulent flow, and passage ways approximately 38 to 50 mm (1.5 to 2 in.) apart and 12.7 mm (0.5 in.) from the cavity surface (Figs. 10.3 and 10.4).

Larger molds may be equipped with several—up to three or more—independent cooling zones (Fig. 10.5).

10.3 Importance of Fast Mold Cooling   153

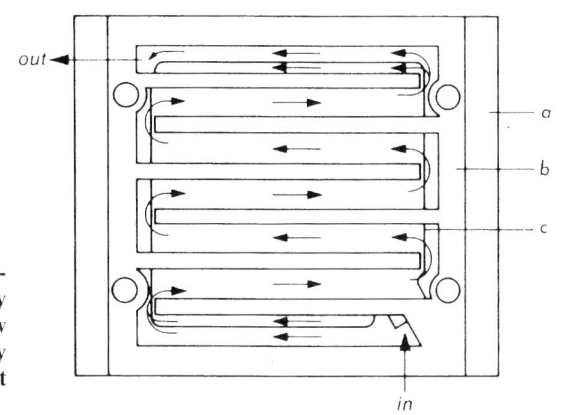

Figure 10.3 Channel system with labyrinth-type water flow produced by baffles. (a) Mounting plate (b) Blow mold (c) Visible contours of mold cavity for a canister with filler connections at one corner (Courtesy of Hoechst A.G.)

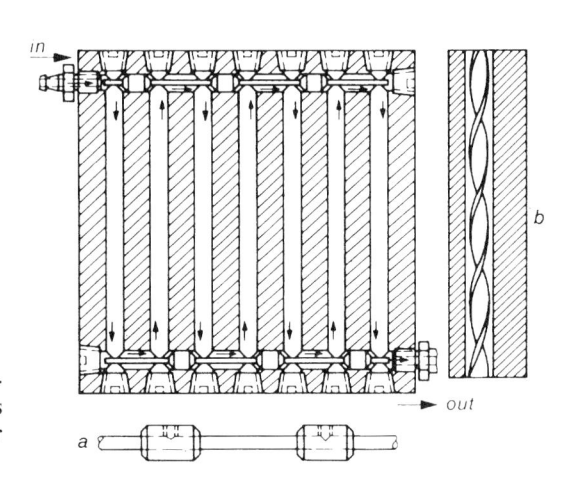

Figure 10.4 Cooling channels with labyrinth-type water flow. (a) Continuous rod with stoppers (b) Inserted copper spiral (Courtesy of Hoechst A.G.)

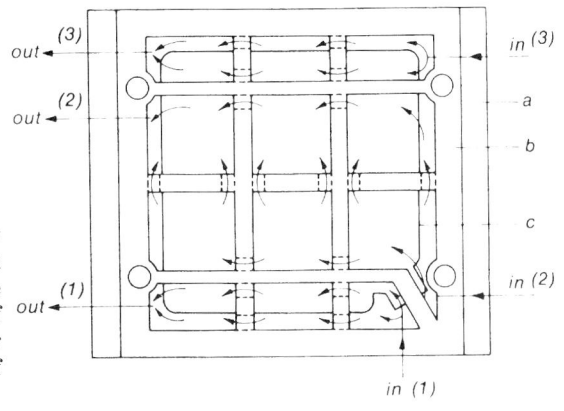

Figure 10.5 Cooling a blow mold by means of three cooling circuits and cooling chambers. (a) Mounting plate (b) Blow mold (c) Visible contours of mold cavity for a canister with a filler connection at one corner (Courtesy of Hoechst A.G.)

10.3.4 Pinch-Off Areas

Generally, around the bottle neck, the pinch-off areas, at the top and bottom of the cavity, have a greater resin mass than other areas. Such areas as well as thicker wall sections, therefore, often require additional cooling. Otherwise, these sections would still be viscous while the thinner wall sections have solidified when the piece is ejected. This may result in a deformed piece or one with nonuniform shrinkage and resulting warpage. That is why even a simple mold has two or more cooling systems for each half.

10.3.5 Blowing Pin

Occasionally, the blowing pin is also cooled. Sometimes, air cooling from the outside is provided for the pinched "tail" of the parison protruding out of the mold bottom. The "tail" is much thicker than the wall and cools correspondingly slower.

10.3.6 Internal Cooling

Air may be circulated inside the blown part to speed up its cooling. Cooling time is strongly affected by the extrusion melt temperature of the blow molding cycle. It has been shown that an increase, or decrease, of $-12.2\,°C$ ($10\,°F$) in melt temperature can extend, or shorten, the cooling cycle by as much as one second.

10.4 Cutting and Welding Parison (Pinch-Off)

Because of the comparatively high pressure and mechanical stress exerted on the mold bottom when (in the closing step) it pinches one end of the parison together, the pinch-off in a nonferrous metal mold is frequently an insert made of hard, tough steel. The effect on the blown part always shows in the so-called weld line.

10.4.1 Pinch-Off Section

The pinch-off section does not cut off the excess parison "tail" (Fig. 10.6). Its protruding edges cut nearly through, creating an airtight closure by pinching the parison along a straight line which makes it easy later to break off or otherwise remove the excess "tail" piece. A high-quality pinch-off of a thick-walled parison is more difficult to obtain than that of a thin-walled parison. However, much depends on the construction of the pinch-off insert.

10.4 Cutting and Welding Parison (Pinch-Off)   155

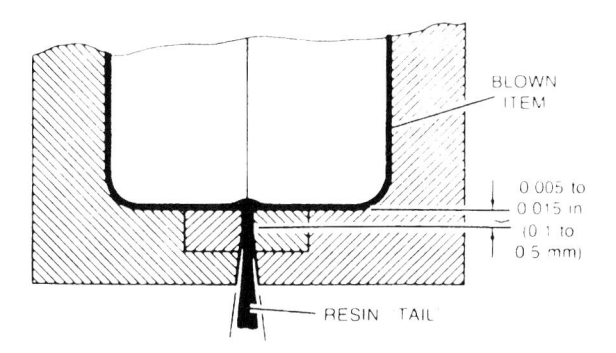

Figure 10.6 Pinch-off tail

The pinch-off should not be knife edged but, according to some molders, should be formed by lands about 0.13 to 0.38 mm (0.005 to 0.015 in.) × 0.5 to 3 mm (0.02 to 0.12 in.) long. The total angle outward from the pinch-off should be acute, up to 15° (Fig. 10.7). These two features combine to create a welding line that is rather smooth on the outside and forms a flat elevated line or a low bead inside, not a groove. A groove, which weakens

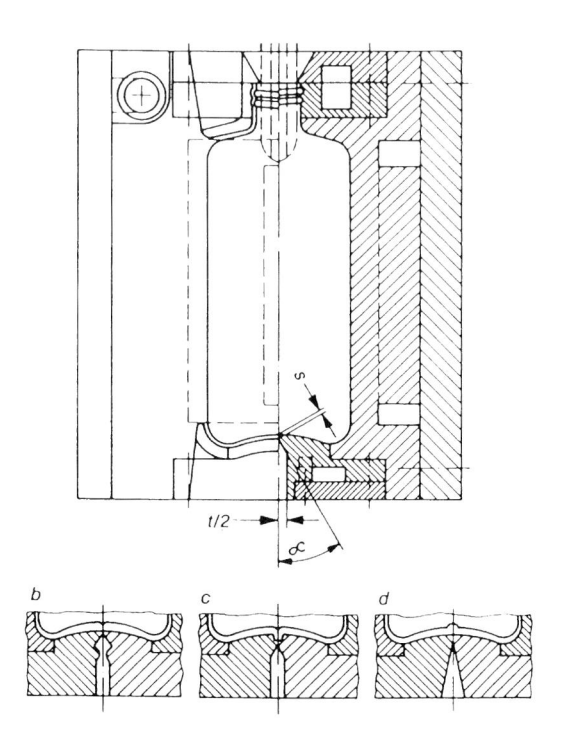

Figure 10.7 Pinch-off configurations (Courtesy of Hoechst A.G.)

156 Understanding the Mold

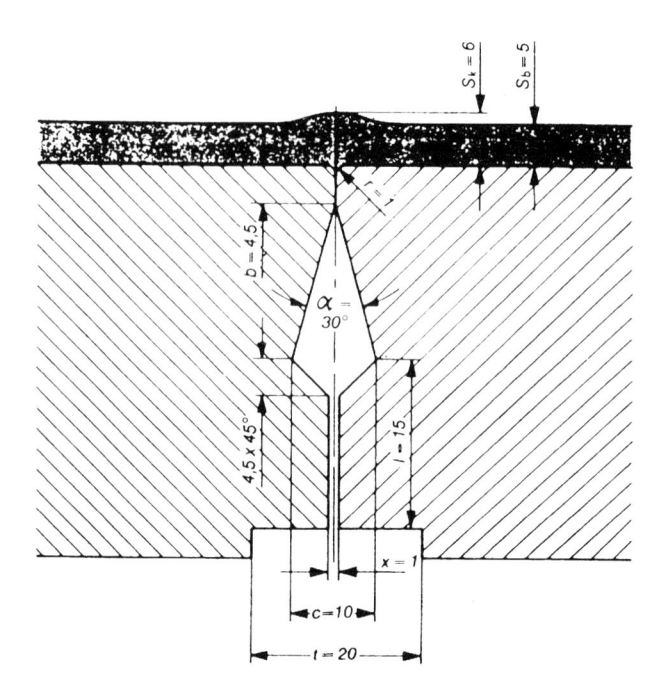

Figure 10.8 Optimized dam for a 60-liter can made from HMW-HDPE (Courtesy of Hoechst A.G.)

the bottom along the seam, may be formed when these two features of the pinch-off are missing.

10.4.2 Uniform Weld Lines

One method of obtaining more uniform weld lines is to build "dams" into the mold halves at the parison pinch-off areas. These "dams" force some of the molten resin back into the mold cavities to produce strong, even weld lines. Figures 10.8 and 10.9 show a cross-section through a well-shaped container bottom and one through a poor one due to an incorrectly constructed pinch-off insert.

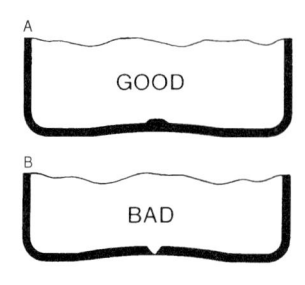

Figure 10.9 A good weld line, obtained by means of a well constructed pinch-off such as shown in b. A grooved and grooved consequently, weal weld line, owing its shape to a knife-edged pinch-off and too wide of too narrow a relief angle (Courtesy of Equistar Chemicals, LP)

10.5 High-Quality, Undamaged Mold Cavity Finish

High-quality mold cavity finish and undamaged inside surfaces are essential in polyethylene blow molding to avert surface imperfections in the end product. If the highest possible gloss of the end product is desired, the mold cavity should be sand blasted with 100-grit flint sand and have vacuum assists for removal of entrapped air. If other end product finishes are desired, the mold cavity should be finished accordingly.

Even a first-class machining job inside the mold cavity cannot prevent the occurrence of parting lines, especially if the blown item has a very thin wall.

10.6 Effects of Air and Moisture Trapped in the Mold—Venting

Air may be trapped between the mold walls and the hot, still soft piece, marring the surface of the piece. This will happen especially when thick-walled, large pieces are blown. In such

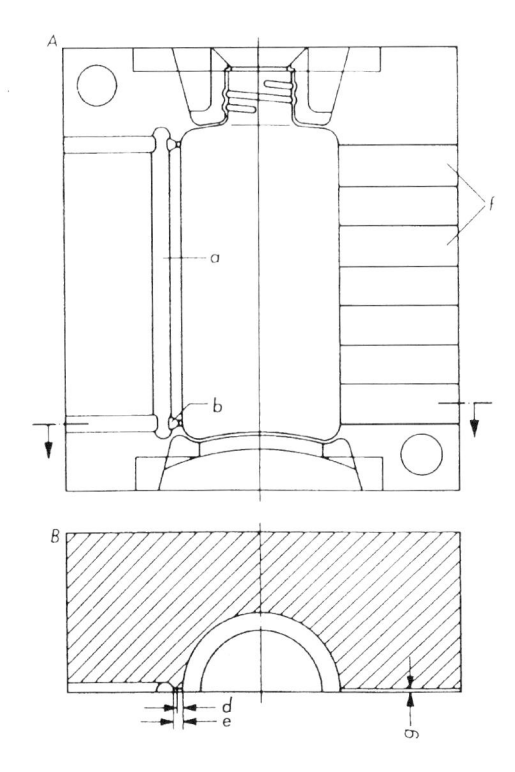

Figure 10.10 Part line venting of mold cavities in the mold parting line. (A) Facilitating free escape by means of a groove at a distance from the cavity edge; venting in the base and shoulder edges by funnel-shaped venting channels b; d = 0.5 mm (B) Venting slits in the parting line f (0.1 mm deep 0.25 mm wide); g = 0.1 mm (Courtesy of Hoechst A.G.)

158 Understanding the Mold

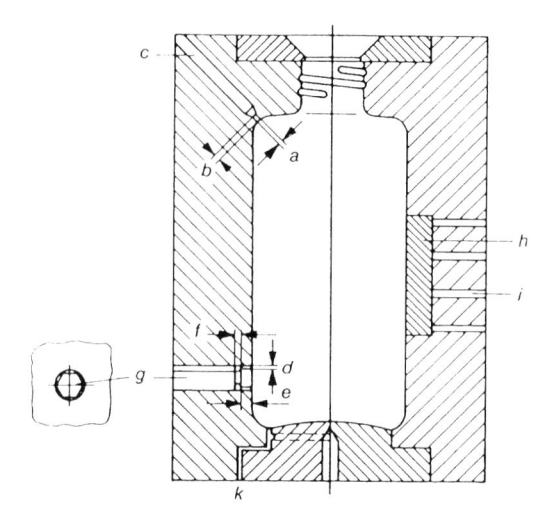

Figure 10.11 Cavity venting suitable methods for mold cavity venting. Vents: a 0.1–0.3 mm diameter, b 0.5–1.5 mm, c blind hole, d camber height of the free circular section 0.1–0.2 mm, e slit depth 0.5–1.5 mm. F angular groove, g bolt with venting slit, k ventilation channel in the fitting point of an insert. Mold cavity venting by means of plugs and plates made of sintered metal, h plate made of sintered metal, I vent (Courtesy of Hoechst A.G.)

cases, the mold must be vented by either sandblasting—resulting in a matte outer surface of the piece—or by grooves in the separation lines or by vents in the mold [3–5].

Generally, about one half of the parting line periphery is vented to a depth of 0.05 to 0.1 mm (0.002 to 0.004 in.). In venting difficult areas, such as handles or thread inserts, holes are usually drilled into these areas so that they vent to the atmosphere. The vent holes are normally about 0.2 to 0.3 mm (0.008 to 0.010 in.) in diameter. Particular care must be taken when drilling these holes so that the mold cooling cavity is not pierced (Figs. 10.10 and 10.11).

Moisture in the blowing air may result in marks on the inside of the blown part. This may result in high reject rates, especially if the part is transparent or translucent. Moisture in the blowing air can be removed by means of a heat exchanger which cools the compressed air, or by traps and separators in the pipe lines.

10.7 Injection of the Blowing Air

Injection of blowing air can be done by various means, such as downward through the core, or through a blowing needle inserted sideways through the mold wall, or from below through a blowing pin moved up into that end of the parison that will become the (frequently threaded) neck of a bottle.

Sometimes, different blowing devices are used in combination. Like every step in the blow molding cycle, blowing time and duration must be well coordinated with all other parts of the cycle. As explained before, compared with the cooling time, blowing time is very short.

To obtain rapid inflation of the hollow piece, the volume of injected blowing air should be as large as possible. The opening through which the air enters the mold must, of course, be adequate.

The thinner the wall thickness and the lower the melt and mold temperatures, the faster the blowing rate should be and the higher the blowing pressure, up to about $10.5\,kg/cm^2$ ($150\,lb./in.^2$) for very cold molds and thin-walled parts. (Compressed air injected into a cold mold may lose some pressure because a cooling gas contracts.) High blowing pressure requires correspondingly high clamp pressure to keep the mold tightly closed during the blowing step.

10.8 Ejection of the Piece from the Mold

Ejection of the blown piece can be ejected forward between the mold halves or downward, provided the press is built in such a way that a free fall from between the open mold halves is possible. Many machines have an automatic ejector, or stripper, assembly. Ejector, or knockout, pins or plates push the piece out with a rapid motion, preferably hitting it on a trim area so that the piece itself will not be distorted. Ejection of the blown piece is part of the automatic blowing cycle. If the piece is not too large, it can be blown forward out of the mold by an air jet from behind. Very large pieces are sometimes manually removed from the mold by the operator. To reach it, the operator must first push the protective gate or shield aside, which automatically stops every mold or platen movement so that the operator's hands are in no danger of being hurt between the mold halves. Mechanical pickers are a preferred method.

Automatic stripping, being a very fast operation, requires a pneumatic (air) pressure system, such as the blowing cycle, cut-off at the die, operation of automatic valves in the die head, etc. Because so many moving parts are thus actuated by air, it is sometimes recommended to oil the air in the pneumatic system lightly.

10.9 Bottle Molds [2]

10.9.1 Neck Ring and Blow Pin Design (Figs. 10.12 to 10.17)

The neck ring and blow pin work together to form the container's finish. The finish, the area of the container where the closure will be applied, is the relationship of the threads to the top sealing surface and the shoulder of the container.

The neck ring is an aluminum block mounted on the top of the main body. Within the limits of the container design, any of several styles of finish can be interchanged. For

160   Understanding the Mold

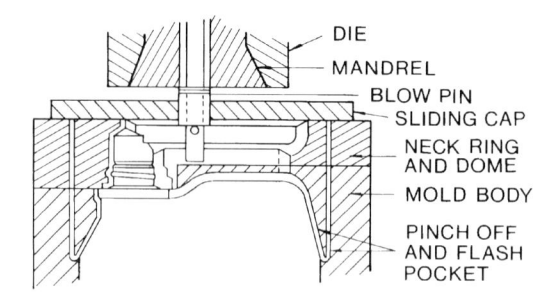

Figure 10.12 The guillotine-and-face method of finishing is used for handled and flat oval containers (Courtesy of Johnson Controls)

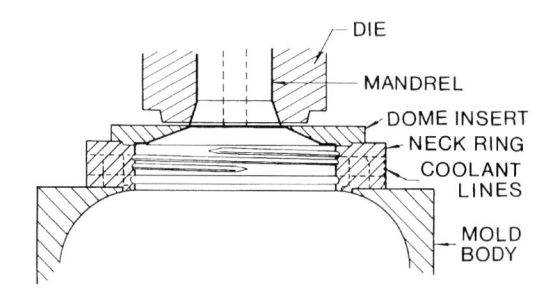

Figure 10.13 The fly-cut method is used for wide mouths where the parison falls inside the neck (Courtesy of Johnson Controls)

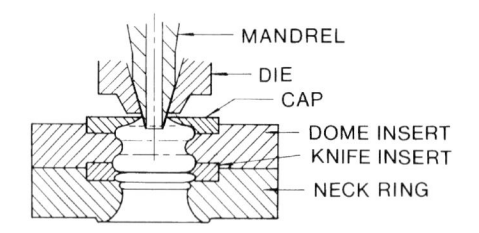

Figure 10.14 The spin-off method finishes other containers where the parison falls inside (Courtesy of Johnson Controls)

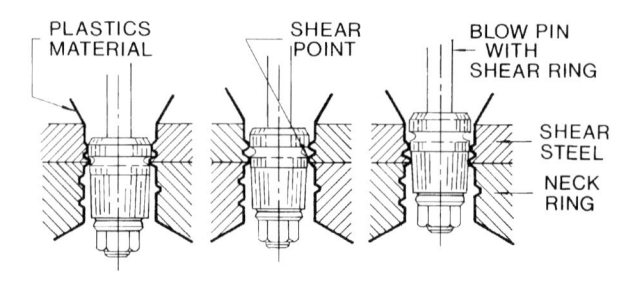

Figure 10.15 In pull-up operation, raising of the blow pin shears the material to achieve a good prefinish (Courtesy of Johnson Controls)

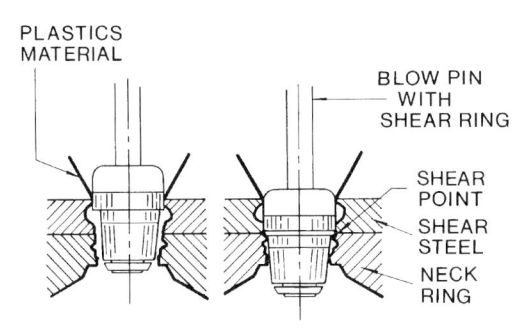

Figure 10.16 The ram-down interference method of achieving prefinish drops the blow pin in the closed mold to shear the material (Courtesy of Johnson Controls)

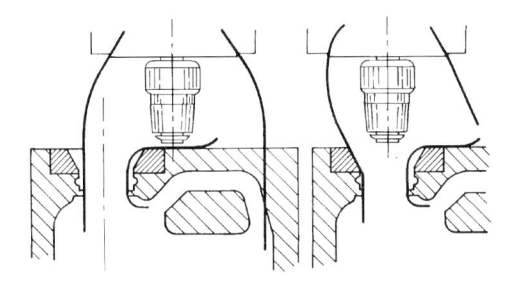

Figure 10.17 In the ram-downside-shift method the mold is closed (left), and the platen shifts to align the blow pin before it is dropped (Courtesy of Johnson Controls)

proper cooling, water lines in the neck ring interconnect with water lines in the mold body (Fig. 10.13).

With the exception of cam and rachet locks used on some specially closures, most of the finish in the neck ring is cut on a lathe. The shrinkage applied to the neck ring carefully calculated, based on the finishing technique and product wall thickness. The shrinkage rates are often different from the applied to the cavity.

10.9.2 Dome Systems

In dome systems, the blow pin is not as critical, and in some applications it is not required. In prefinished systems, the blow pin has an important role in molding the finish. With the dome system, the neck ring forms only the threads of the finish. The blow pin, if required, is used to seal and channel air inside the parison. After molding, the dome top is cut off and the container finish sized and trimmed to specification.

Three basic types of dome neck rings can be chosen, depending on the type of container and the type of trimming equipment. The guillotine–and–face method requires a blow pin and is used for handles and flat oval containers requiring the parison to flash outside the neck. After molding, the trimmer will guillotine the dome top from the container, the pour lip and inside diameter of the finish part would be facing the cutter.

The fly-cut method is used for wide-mouth containers where the parison will fall inside the neck, a blow pin is not required. The parison is sealed against the mold against the die of the head cooling. The dome is removed when the fly-cutter sizes the inside diameter of the finish.

The spin-off method is used for any container where the parison will fall inside the neck. A blow pin is not necessary. A special dome, shaped like a pulley, is molded above the finish. After part trimming, a V-belt fits into the dome, spinning the container against a stationary knife blade that cuts through the dome top in the sharp groove at the top of the finish. Spin-off technology is patented by Johnson Controls, Manchester, MI.

10.9.3 Prefinished System

With the prefinished system, the neck ring and blow pin work together to mold and size the finish to specification before the container is removed from the mold cavity. Some prefinished neck rings require the parison to be extruded over the blow pin and flash must be outside the neck. Choice of the pull-up or ram-down method is dependent upon the type of closure used.

The pull-up method (patented) is used primarily on dairy container closures, which usually seal on the inside diameter of the finish. With the pull-up system, the blow pin moves upward just prior to the mold opening, shearing plastics material to form the finish inside diameter. The diameter of the hardened shear steel mounted on the top of the neck ring and the hardened shear ring mounted to the blow pin are held to a very close tolerance. This ensures the finish inside diameter will be accurate, smooth, and free of burrs.

The ram-down method (patented) is used in containers with closures requiring a flat, smooth, top sealing surface on the finish. Immediately after mold closing, the blow pin moves downward, pushing material into the finish. The stroke of the blow pin is adjusted so that the shear ring mounted on the blow pin contracts the shear steel mounted in the neck ring without interference, pressure, or stress. The diameters of the hardened shear steel and shear ring are also held to very close tolerances, which ensures that the container finish is held within specification. Containers with off-center necks, with the finish offset from centerline, can also use this prefinish method. The sequence of operation is the same except the mold side shifts after it closes and before the blow pin downstroke.

10.9.4 Unusal Problems

Several special features have been developed by moldmakers to deal with unusual problems—side cores for deep undercuts, needle blow pins for container/cover combinations, and provisions for sealed blowing where the container is sealed before the mold is opened to maintain sterile conditions inside.

10.9 Bottle Molds   **163**

Two other special feature developments are the "rotating mold" and the "sliding bottom" (both patented).

The rotating mold was developed for container designs that have finish centerlines at some angle to the container centerline. The system permits use of the ram down prefinish method (Fig. 10.18).

The sliding bottom is a feature that prepinches and seals the bottom of the parison, permitting it to be puffed up like a balloon. This device allows use of a smaller diameter parison than normal. The smaller parison pinch, which results at the bottom, is of value on certain containers filled with a stress-crack agent.

The most common spacial feature is "quick-change" volume-control insert. Rigid volume control is a must for dairy containers. Unfortunately, HDPE containers shrink slowly and change size for many hours after molding. Because of production volume and control, many dairies have switched to fill containers molded several days earlier. Volume-control inserts, which displace the difference in size, are added to mold to ensure that volume and fill levels are the same in both containers at the time of filling. The device works because shrinking is reduced to virtually zero for the life of the HDPE container when it is filled with milk or juice and stored at cold temperatures.

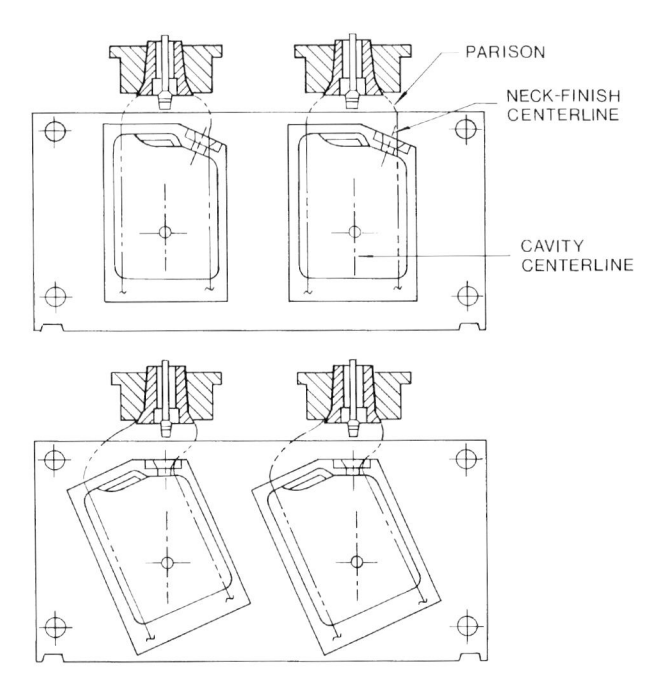

Rotating molds (shown open at the top and closed below) are used for containers with finish centerlines at an angle to the container centerline.

Figure 10.18 Rotating molds (shown open at the top and closed below) are used for containers with finish centerlines at an angle to the container centerline (Courtesy of Johnson Controls)

10.10 Injection Blow Molds

The injection blow molding process has:

- Preform mold, which consists of the preform cavity, including neck ring, insert, and core rod assembly;
- Blow mold consisting of bottle cavity, neck ring insert, and bottom plug insert;
- Manifold;
- Nozzle; and
- Stripper.

See Fig. 10.19.

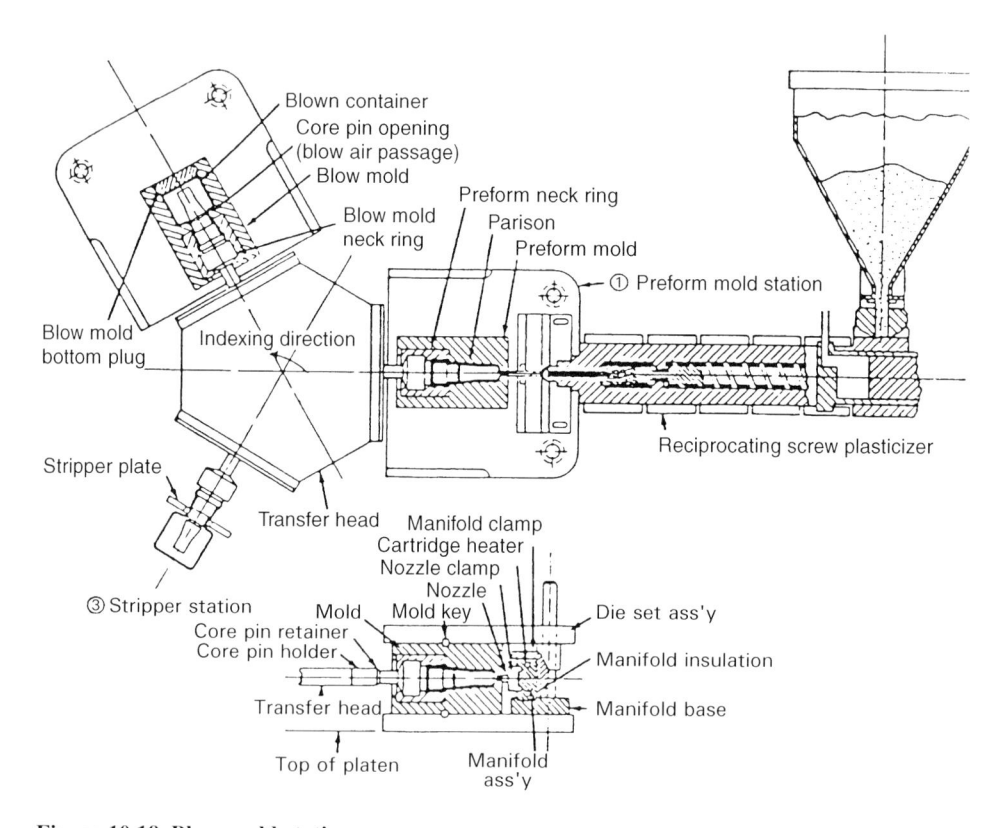

Figure 10.19 Blow mold station

10.10.1 Parison (Preform) Mold (Fig. 10.20)

Four basic rules govern this preform cavity design. The first concern is the core. The rod length-to-diameter ratio approximates 10:1 or less. This ratio is usually figured on the

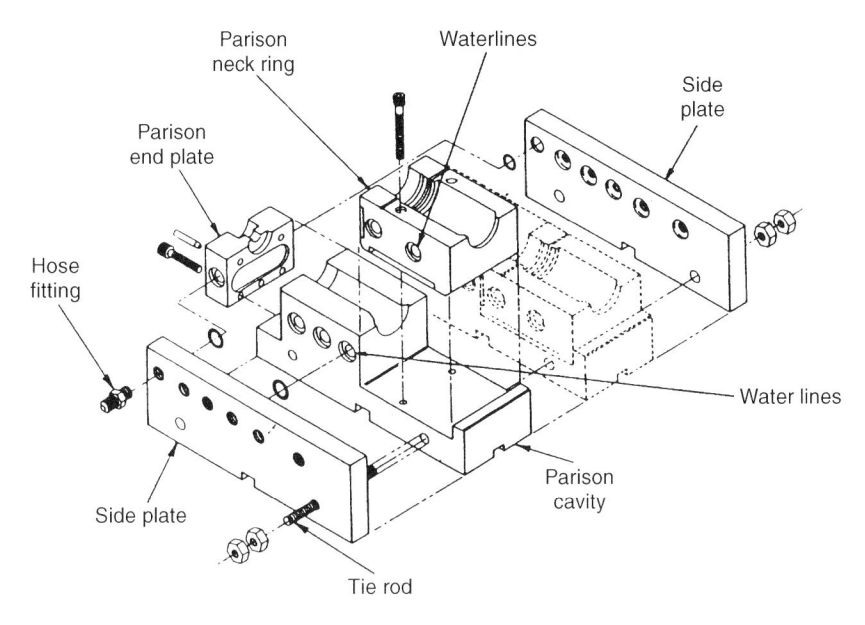

Figure 10.20 Parison mold assembly (Courtesy of Johnson Controls)

overall height to neck finish diameter of the bottle. The second rule concerns the ratio of preform size to maximum bottle size, which ideally is 3:1 or less. More often it takes the maximum bottle diameter width or depth and the neck finish diameter. The maintenance of this rule ensures uniform and consistent bottle cross-sectional wall distribution. Finally the third rule regards the parison wall thickness. The ideal is between 2 and 5 mm. A thicker wall than 6 mm causes a difficult to control temperature condition and may act unpredictability during expansion. A thinner wall of 2 mm thick may also act unpredictably.

The injection blow mold advantage is the diametrical and longitudinal programming of the parison shaping the parison mold cavity or core rod or both. The fourth rule, which is particularly important in oval bottles, is that in an angular cross-section, the heaviest area should not be more than 30% thicker than the lightest area. The shaping is done in the cavity with the core being round. With a higher ratio, the selective fill of material during the injection phase creates a vertical weld line in the bottle. Avoiding condition in turn restricts the bottle ovality to 2:1. Thus the width does not exceed twice the depth. With multicavity arrangements, each parison cavity has an injection nozzle of decreasing size, ensuring balanced even material flow through the injection manifold.

10.10.2 Neck Ring Insert

This part has four main functions:

1. It forms the finish or threaded neck section.

2. As it is an insert, it provides a relatively low cost and thus easy method to change the size or style of the finish.
3. It holds and centers the core rod in the cavity.
4. It provides thermoisolation.

The neck finish area of the parison is cooled to 5 °C (41.04 °F) to prevent distortion of its shape; the rest is kept at a temperature of 65 °C and 135 °C (149 °F and 275.04 °F). The water lines are usually drilled close together, perpendicular to cavity axis in both cavity and neck ring. The water flows from one cavity to the next.

10.10.3 The Core Rod Assembly (Fig. 10.21)

This assembly also has four functions:

1. It forms the interior of the preform.
2. It provides support for the parison of the bottle during transfer.
3. It supplies the valve where air enters to expand the parison. Note: The valve is in the shoulder area or the tip, depending on the shape of the bottle, such as when a core rod has low length-to-diameter ratio.
4. It has a "blow by" groove. The purpose of this angular groove, which is located near the seating shank, and 0.1 to 0.25 mm deep (0.0039 in. to 0.0098 in.), to seal the parison and prevent excessive air loss during blowing. Thus eliminates elastic reaction of the parison during the transfer between cavities.

10.10.4 Materials for Parison Cavity and Core Rods

The parison cavity is constructed of prehardened tool steel with a hardness of 31–35 HRC for polyolefins; for rigid resins, the parison cavity is made of A.2 steel, which is air hardened to 52–54 HRC. The neck insert for most resins is more from A-2 tool steel. For extra strength the core rod is made from L-6 tool steel, hardened to 52–54 HRC. All the cavity surfaces are highly polished and chromium-plated with the exception of the neck ring insert for polyolefin resins, which is occasionally sand blasted with No. 120 grit.

For bottle cavities and neck rings, aluminum steel or beryllium–copper is used. For polyolefin resins, No. 7075 aluminum is used. For venting trapped air the surface is usually finished to a No. 120 grit sand blast. For rigid resins an A-2 tool steel air-hardened to 52–54 HRC is used. The surface finish is highly polished chromium plate. Cast beryllium–copper is often used for minute detail. The water lines, as with the parison cavity, are drilled as close together as possible, perpendicular to the cavity axis.

10.10.5 Design Details of the Blow Mold Cavity (Fig. 10.22)

The final shape of the bottle is defined by the cavity, the only design constraint being that the width of the cavity should not exceed the depth by twice the depth to compensate for

10.10 Injection Blow Molds   167

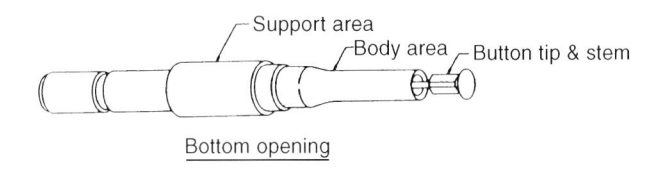

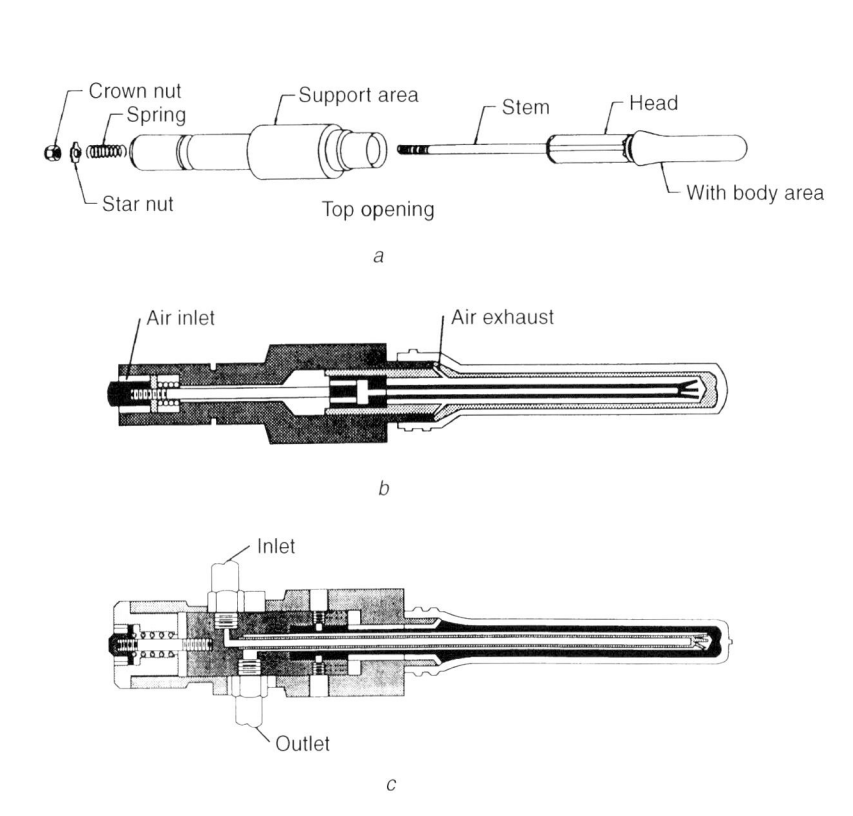

Figure 10.21 Core Rods (a) Typical assemblies (b) Air-cooled rod (c) Liquid-and air-cooled core rod (Courtesy of Johnson Controls)

resin shrinkage after molding cavity sizes are slightly enlarged. Shrinkage rates vary with resin and prevailing process conditions. For polyolefin shrinkage varies between 1.6% and 2%; for ridged resins 0.5% is used. For the neck finish dimension a higher rate is used because of heavy sections.

10.10.6 Vents

Vents are placed along the mold parting surface to allow escape of trapped air between the expanding parison and the cavity. This should not be too deep because an objectionable

168 Understanding the Mold

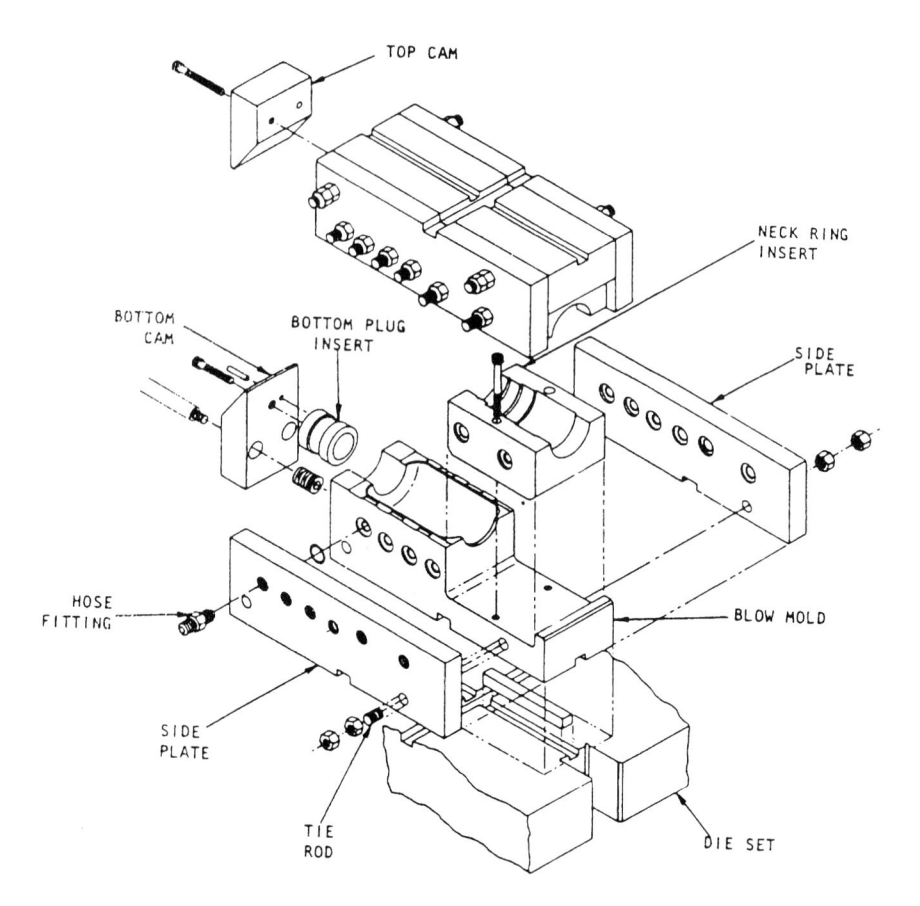

Figure 10.22 Blow mold assembly (Courtesy of Johnson Controls)

mark would be left on the bottle. Note that the air pressure of 1 MPa (145 psi) is used in injection molding. Thus the vents should not exceed 0.05 mm (0.002 in.) deep.

10.10.7 Neck Ring Insert

This insert is used in the bottle cavity in a similar way as the parison cavity. The thread diameter dimensions in the bottle cavity are 0.05 to 0.25 mm (0.002 in. to 0.0098 in.) larger than the parison cavity. The bottle neck ring, unlike the parison, does not form the finish detail. It serves only to secure the already formed neck.

10.10.8 Bottom Plug Insert

The bottom or pushup area of the container is formed by the bottom plug insert. In some molds this insert must be retractable. In general, the pushup of polyolefin bottles can be stripped without side action, particularly if the height is less than 5 mm (0.196 in.). With ridged resins, this height is reduced to 0.8 mm (0.025 in.). An air cylinder cam or spring mechanism is used when side action is required.

10.10.9 Die Sets

The parison and bottle molds are mounted onto a die set. This is then mounted onto the platens of the injection blow mold. To position the die sets are located by key sites in two directions on the upper and lower cavities. Precise alignment between plates is provided by guide pins and bushings (Figs. 10.23 and 10.24).

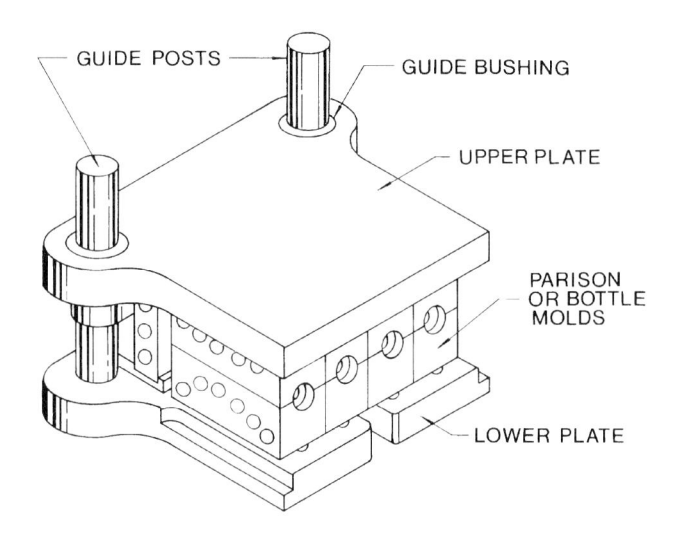

Figure 10.23 Die set for maintaining position and alignment of injection blow mold cavities

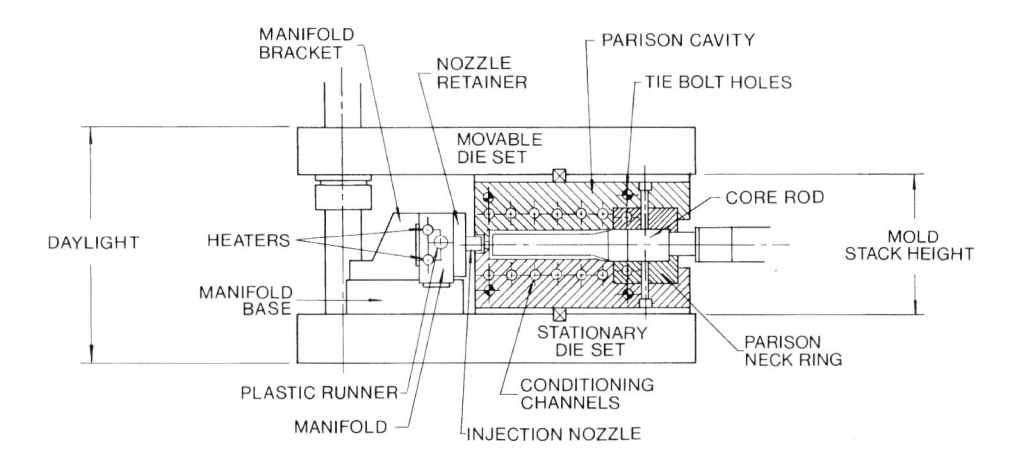

Figure 10.24 Die set with tooling and nomenclature

10.10.10 Injection Blow Mold Tooling Summary

Injection blow molding tooling must be engineered to precise tolerances, molding dimensions to $+/-0.015$ mm. Core rods must be located closely front to back, and left to right of centering of the parison and bottle cavities. When too tight, the mold could be damaged by binding. If too loose, the resin could flash around the shank area of the core rod, or the rod could shift sideways, causing an uneven wall. Also, the parts must be interchangeable between assemblies and cavities. Thus, a critical factor is the need for precision in injection mold tooling.

10.11 Conclusion

Sound product design is the first step in mold design. The moldmaker is the last step. The moldmaker is a highly skilled craftsman whose judgment and ability are used to duplicate the cavities, hand finish and fit the parts, engrave the lettering, and manufacture the molds. His commitment, attention to detail, and pride are as important as the design itself. If the molds are maintained they will be troublefree and give long service (see Figs. 10.25 and 10.26).

Check lists for ordering a mold are given in Figs. 10.27 and 10.28.

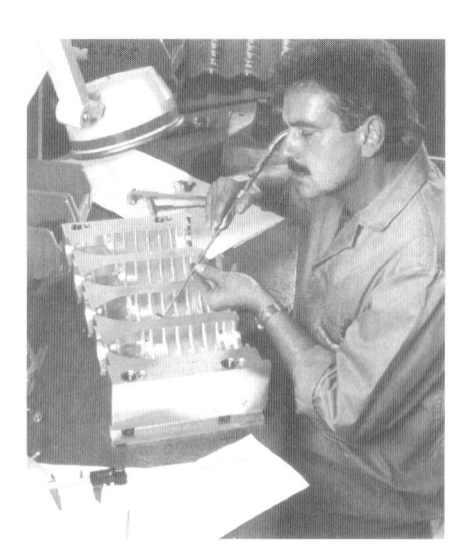

Figure 10.25 Mold maker completing an injection stretch blow mold for processing PET (Courtesy of Krupp Kautex)

Figure 10.26 Three axis measuring machine with equipment for data transfer to the CAD system (Courtesy of Krupp Kautex)

```
Date_____
Company_____
Part Name_____
Part No._____
Molding Press_____
Shut Height (min)_____Length x Wedth (max)_____
Platen Layout
Plastic Shrink_____"/"____Model Furnished: Yes_____No_____
No. of Cavities_____Mold Sections_____Knife Insert_____
Type: Cast Alum._____Mach. Alum._____Other_____
Cooling: Drilled_____Channel Back_____Cast Tubes_____
Type Blow: Top_____Bot._____Blow Pin Size_____Dia._____
             Air Cyl._____Needle Size_____
Pinch off: 100%_____Ends only_____Inserted_____H$_2$O_____
          Inserts: Alum._____Steel_____Bronze_____
Core Pull: Mech._____Cyl._____Air_____Hyd._____
Prepinch:_____Teflon Coated_____Spring_____Cyl._____
                   Water Cooled_____
Tread Insert: Info_____Steel_____Alum._____H$_2$O_____
Engraving: Cast in_____Inserted_____Removable_____
           Interchangeable_____Dial-A-Date_____
Cavity Finish: Text Info._____Polished_____Blasted_____Grit_____
Riser System: Rails_____I-Beam_____Alum._____Steel_____
Eject. System: Cyl. Act._____Spring Act._____Location_____
Mold Mounting: Flanged_____Tapped Holes (size & Layout)_____
Mold Identification:_____Trade Mark_____
Cavity No._____Part No._____Location_____
Venting: Drilled Holes/Size_____Core Vents_____Other_____
```

Figure 10.27 Extrusion mold check list

```
Date_____
Company_____
Part Name_____
Part No._____
Molding Press_____
Shut Height (min)_____Length x Wedth (max)_____
Nozzle Size_____
Heat Isolation at neck ring or holding dia._____
Neck ring material_____Neck finish_____
Platen layout_____
Plastic Shrink_____Model Furnished_____
Number of cavities_____
Parison/Core rod specifications_____
Eject. System: Cyl. Act._____Spring Act._____Location_____
Stripper Plate_____
Mold Identification:_____Trade Mark_____
Cavity No._____Part No._____Location_____
Venting_____Core Vents_____Other_____
```

Figure 10.28 Injection mold check list

10.12 References

1. Raddatz, E. and Gondo, C. *Hoechest Blow Molding Handbook*, Frankfurt, Germany
2. Irwin, C. Johnson Controls, Manchester, MI
3. Suit, M., Blow Molds, In *Modern Plastic Encyclopedia* (1986) McGraw-Hill, New York
4. *A Short Course in Blow Molding High Density Polyethylene* (1970) Chevron Chemical Co., Houston, TX
5. Irwin, C. *Extrusion Blow Molding Tools* (1985) Center for Professional Advancement

11

Computer Aided Design and Engineering for Mold Making

The mold making industry now operates in a highly competitive market. The companies that survive in this environment will be those that offer exceptional service at a reasonable price to their customers. The United States and Europe in particular, with their high wages and living standards, can best compete on the basis of superior technology, service and quality.

In the previous chapters, it has been shown that the emerging technology that analytically eliminates "the art of design" of the part to be molded will emerge as the preferred design system. This is not to say that designs developed through these methods are not artistic, but rather the techniques used are based upon the advanced capabilities of computers, as extensions of the human mind. The high speed of computers allows complex calculations to be made, opening new vistas to the designer/engineer. Those who take advantage of these developing technologies will create better designs, and meet predicted performance criteria. Art and experience are not replaced but enhanced and augmented. Guess work, although based on experience and rules of thumb, is eliminated. The overall resultant design is vastly improved [1].

11.1 Advantages

- Full documentation of complex shapes,
- volumetric analysis without building solid models (Fig. 11.1),
- material flow and rigidity, and
- images of models for marketing presentation and approval.

Databases may be used for programming computer numerical control (CNC) machine tools and/or speed the design through modular or parametric construction [1] (Figs. 11.2 and 11.3).

The purpose here is to show how the mold maker may use this information to achieve a competitive edge.

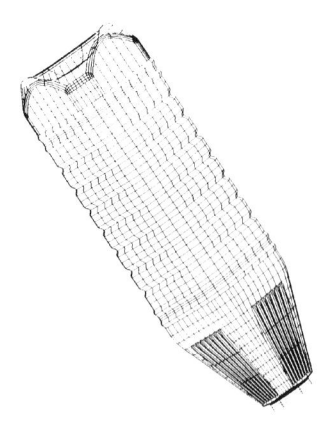

174 Computer Aided Design and Engineering for Mold Making

Figure 11.1 Three-dimensional molding, models are used to calculate the fill volume of the containers (Courtesy of Krupp Kautex)

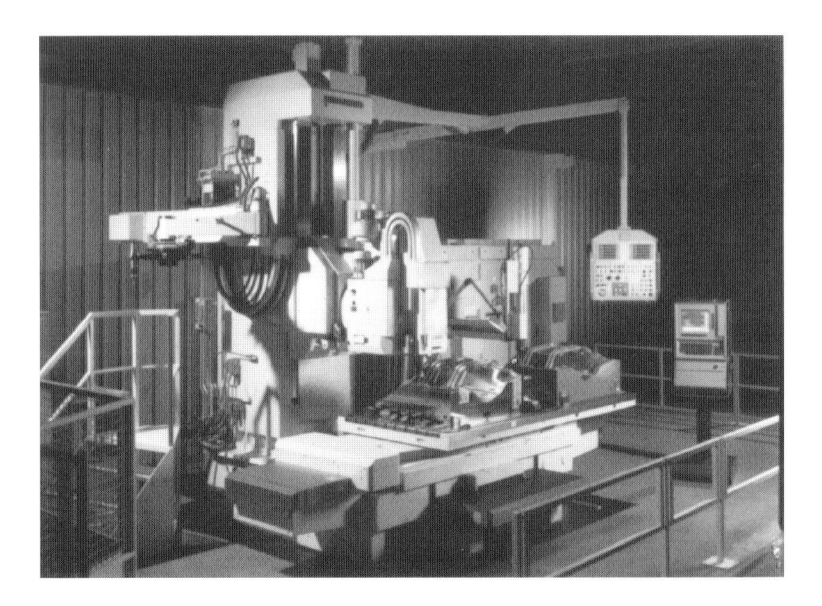

Figure 11.2 CNC copy milling machine (Courtesy of Krupp Kautex)

11.2 Systems and Methods

11.2.1 Analytical Personal Computer

The personal computer has brought CAD and CAE to the desktop. A decade ago, investment in computer technology meant purchase of expensive mainframe and minicomputers, and hiring a specialist to maintain them. Improvements in the "friendliness" of user interface, processing power, speed, and above all cost has brought computers within economic grasp of even the smallest of shops. Systems developed for

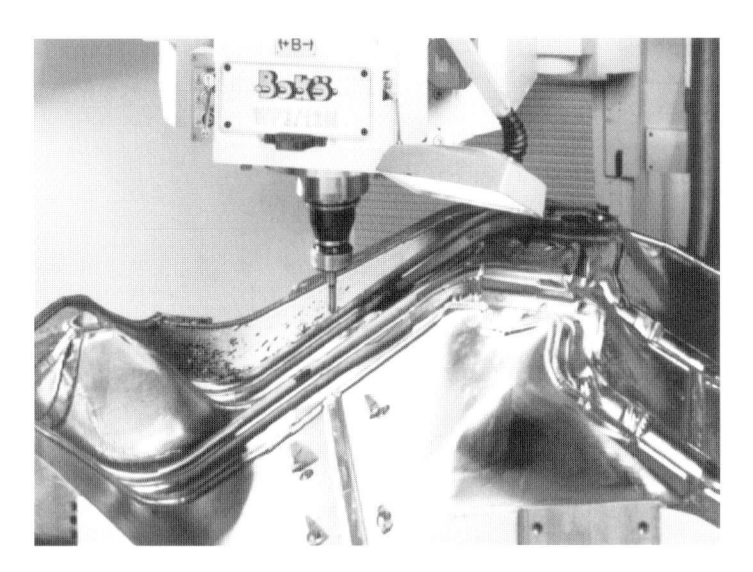

Figure 11.3 Profile finishing on a CNC milling machine (Courtesy of Krupp Kautex)

IBM PC-AT and compatible machines are weaning designers from the drafting board forever [1].

11.2.2 Minicomputer

The minicomputer workstation is the next step up. These stations usually start with fast 32-bit or higher microprocessors. The Unix operating systems with high-resolution 19-inch color monitors have the capability to display thousands of colors simultaneously. They function as general-purpose computers, which perform the primary task such as computer aided design and drafting and evaluation exercises (such as Finite Element Analysis and flow analysis discussed in previous chapters). They may be purchased as a complete package from several vendors as individual units. Specific units can be developed by the user to perform tasks that are proprietary to an organization. An important advantage of individual stations is that they can be networked (interconnected), so that printing, plotting, scanning, libraries, and mass storage facilities can be utilized by a large number of employees in the organization. See Fig. 11.3 for information on manufacturers of stand alone and/or networked work stations for CAD/CAM systems.

11.2.3 Network Station Approach

Large companies with several ongoing projects may install many workstations and have a mainframe that supports all workstations. These mainframe computers support powerful software programs with a fast response. They are very expensive to acquire and to maintain and often become overutilized. When a mainframe system becomes overutilized, the cost of adding additional capacity can be very high, which forces the company to make do with the present, obsolete system. For this reason, many organizations are choosing the network workstation approach for their CAD/CAM resources. Each additional seat may be added and would not affect productivity of others [1].

11.3 Utilizing CAD/CAM in a Mold Making Organization

To be effective in the utilization of the computer the mold making company, department, or group should be organized in all phases from the quotation, placement of order to the completed job.

Organizations ideally have three major departments: Business, Engineering, and Manufacturing (Fig. 11.4). This combination of departments creates the internal context for jobs in the manufacturing environment. The coordination of these departments utilizing the latest technology provides a competitive edge. The ultimate success of the organization demands a thorough knowledge of the technology and a high degree of communication between the various departments and individuals involved, including the customer.*

As a manufacturer of custom products, every order constitutes a job. Every job requires a series of premanufactured parts that are assembled to produce the finished product(s). Finished parts consist of several categories:

- Purchased parts used for direct assembly,
- purchased parts needing additional manufacturing before assembly, and
- raw materials need complete manufacturing for assembly.

Manufacturing activities can be accomplished either internally or externally as a subcontracted process. Assembling the components is the final activity that converts the raw materials into finished products. From a business perspective, every job has certain revenues and expenses (costs). For a job to be profitable, the total revenue must be greater than the total costs.

The outline of the activities and phases for mold building are shown in Fig. 11.5. Some jobs may vary slightly and some may be phased out early. As a result, every job will not pass through every phase.

11.3.1 Engineering Activities

Whenever a job becomes activated, the engineering department becomes responsible for that job. The engineering department must specify all the information needed to properly complete a job. Every job requires the purchase and manufacture of exact quantities and

---

*The material and research for the chapter dealing with application of computers in a mold making company was provided and written by Robert Kennedy of Kennedy Tool & Die, Inc. 225 West Maint Street, Birdsboro, PA.

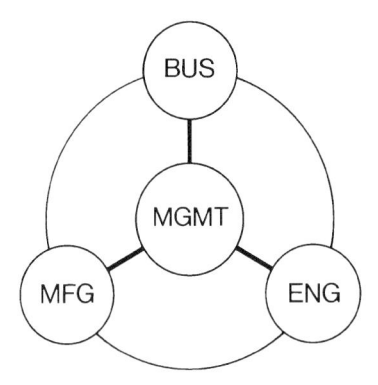

11.3 Utilizing CAD/CAM in a Mold Making Organization

Internal Context Of Departmental Organization

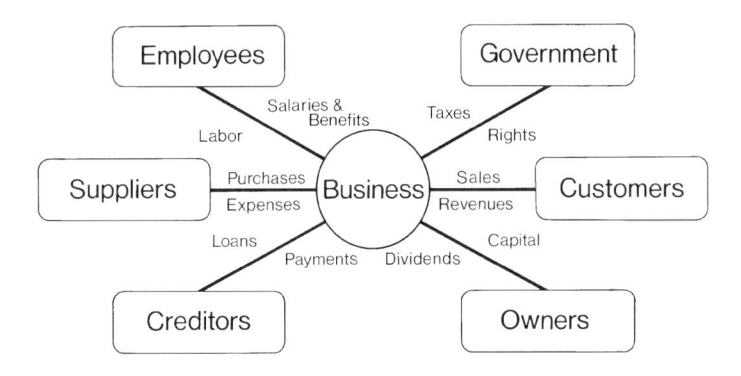

External Financial Context Of The Company

Figure 11.4 Mold making company organization (Courtesy of Kennedy Tool & Die)

types of components which must be manufactured and assembled into the finished products. A complete engineering package for a job includes the following items:

- Part drawings: The part drawings provide a graphical representation of every part requiring manufacturing and/or assembly. This includes all relevant numeric dimensions, tolerances, manufacturing notes, revisions, and bill of materials identification. All parts must be manufactured to the part drawing specifications.
- Bill of Materials: The bill of materials provides a complete list of all the raw and purchased materials needed for a job. Using the bill of materials, all necessary materials acquisition and preparation can be tracked and managed. Tracking information should include items such as vendors, date ordered, who ordered, date received, who received, and current status/location.

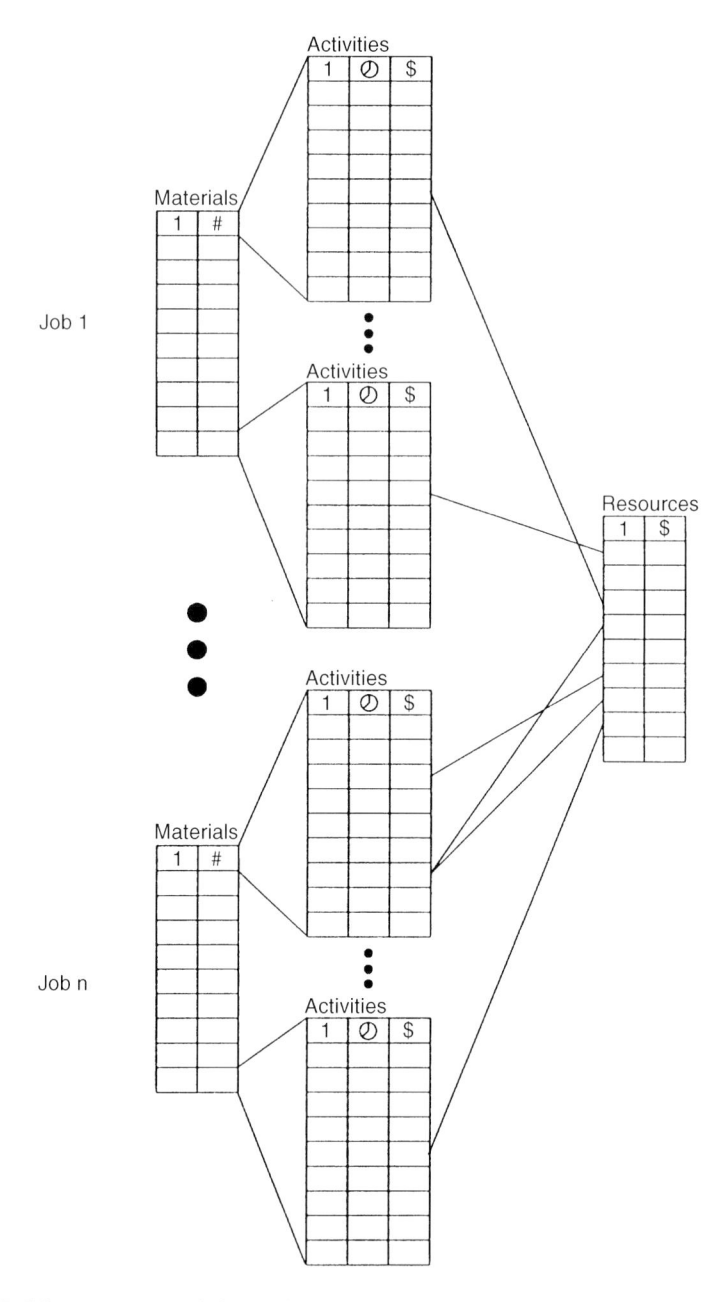

Figure 11.5 Job management information relationships

- Production Plan: The production plan provides a series of operations list. Each list specifies the necessary operations to manufacture a given piece of raw material into a finished part. The type of operation, manual or CNC, is also specified. Using the production plan, raw materials can be efficiently routed, scheduled, and tracked through the various manufacturing operations.
- Inspection Data: The inspection data may take several forms: a list, notes on the part drawing, or a CMM program generated directly from the CAD system. In general, inspection data specify certain dimensions and tolerances which must be checked before a part is approved. This is determined with the customer during part design.

11.3.2 Manufacturing Activities

Once the engineering package has been completed and approved, the manufacturing department becomes responsible for producing the necessary components. Utilizing the information provided by the engineering department, the raw materials for a job can be efficiently manufactured and assembled into finish products. Before manufacturing operations can take place for a job, the following activities must take place:

- CNC Programs: The CNC programs specify the machine control instructions for every practical operation. Practical operation means that certain operations are best done manually, that is, assembly. The CNC programs generated by engineering should be stored in generic cutter location format. This allows operations to be grouped together and posting to proceed for a particular machine control in a single setup.
- Part Routing: Part routing determines the order and grouping of operations. An operation group, one or more operations, is then assigned to a resource. A resource is one or more equivalent man and/or machine combination. The goal of routing is to determine the least setups needed to manufacture a part. This minimizes part handling and maximizes resources utilization. Once operations and resources have been determined, the CNC programs can be organized and posting processed for each setup.
- Part Scheduling: Part scheduling, in general, determines the most expedient method for manufacturing all the parts needed to deliver jobs on time. Scheduling considers all the jobs active in manufacturing at any given time. The scheduling goal is to complete the necessary operations by assigning the appropriate resource on all parts needed for a job, based on a specific deadline. Job progress can be tracked using the production plan. Manufacturing bottlenecks can be identified by considering the routing queue for a particular resource, estimated time for operations completion, and remaining time till expected or requested delivery (Fig. 11.6).
- Quality Assurance: Quality assurance determines that every part for every job has been manufactured to the degree necessary to satisfy the customer's requirements. Quality assurance is achieved by a system that utilizes the inspection data provided by engineering, combined with gauges, manual and automated inspection equipment, and visual methods. Assuming that every part has been manufactured to specification, quality assurance is really quality verification.

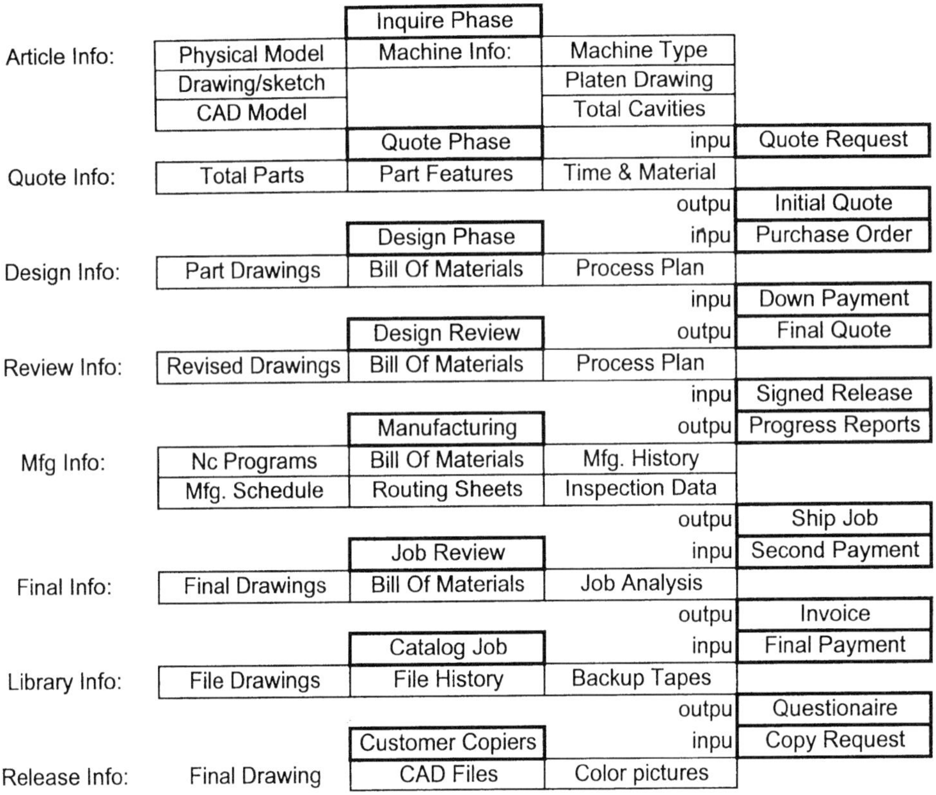

180   Computer Aided Design and Engineering for Mold Making

Article Info:	Physical Model	Inquire Phase	Machine Type	
	Physical Model	Machine Info:	Machine Type	
	Drawing/sketch		Platen Drawing	
	CAD Model		Total Cavities	
Quote Info:		Quote Phase	inpu	Quote Request
	Total Parts	Part Features	Time & Material	
			outpu	Initial Quote
Design Info:		Design Phase	inpu	Purchase Order
	Part Drawings	Bill Of Materials	Process Plan	
			inpu	Down Payment
Review Info:		Design Review	outpu	Final Quote
	Revised Drawings	Bill Of Materials	Process Plan	
			inpu	Signed Release
Mfg Info:		Manufacturing	outpu	Progress Reports
	Nc Programs	Bill Of Materials	Mfg. History	
	Mfg. Schedule	Routing Sheets	Inspection Data	
			outpu	Ship Job
Final Info:		Job Review	inpu	Second Payment
	Final Drawings	Bill Of Materials	Job Analysis	
			outpu	Invoice
Library Info:		Catalog Job	inpu	Final Payment
	File Drawings	File History	Backup Tapes	
			outpu	Questionaire
Release Info:		Customer Copiers	inpu	Copy Request
	Final Drawing	CAD Files	Color pictures	

Figure 11.6 Information flow chart (Courtesy of Kennedy Tool & Die)

11.4 Reference

1. Lee, N. *Plastic Blow Molding Handbook*, Tod Eberle Computer Container/Preform Design

12
Polymers and Plastic Materials

12.1 Basic Polymer Chemistry

A polymer is a linking of many chemical units together through chemical bonds to form what is often referred to as a plastic. The various chemical reactions occur by taking various ingredients (molecules and elements) and arranging them together in different, desirable combinations. This can be done by animal and plant life forms, through random or chance occurrences in nature in a wide variety of ways. Polymers that are formed by nature or natural processes, because they degrade more easily (biodegradation), are often thought to be the most environmentally friendly because manmade energy sources are not utilized. The use of engineered chemical processes to produce polymers is called polymer synthesis. In German, *Kunst* means manmade or art. *Kunststoffe* is literally stuff made by man. This may best describe many polymers.

This chapter will try to help the designer become familiar with some of the terms and concepts involved in working with polymeric materials. Polymers can be classified and understood by relating their properties to use and application. Those polymers that can deform and soften under heat and pressure, after melting, are called "thermoplastic." Those that initially deform under heat and/or pressure, but undergo a chemical change or reaction giving them a new chemical structure are called "thermosetting." This is often the first major difference accorded to the function and use of polymers. The term plastic is a mechanical term applied to how materials deform. Since many polymers do deform easily, the broad and often inaccurate term "plastic" is often applied and will be used here despite the inaccuracy of the generality.

Designers have a variety of plastic materials from which to select and they usually will work with many during their careers. Even those who are familiar with a particular class of material will encounter unexpected processing and/or performance problems from a new or unfamiliar grade [2].

12.1.1 Structure of Matter

Matter is made up of very small particles called atoms. They are the smallest division of matter that undergo classical chemical change. Atoms of the same chemical nature (proton and electron configuration) comprise elements. Atoms of similar chemical nature that have different weights (variable neutron count) are called isotopes. An example is the element carbon. It has several isotopic weights (i.e., 12, 13, and 14 or 6, 7, or 8 neutrons). This is the reason the atomic weight of carbon is 12.011, because it is a matched average of the 12, 13, and 14 mass types. Different elements can join together to form other substances with unique and different properties. For example, two atoms of hydrogen plus one atom of oxygen combine to produce a water molecule.

A molecule is a distinct substance formed by the joining of two or more elements. On a more macroscopic level, different atoms and molecules can combine to form compounds with defined atomic ratios. These may have the same formula but different weights owing to the isotopes found in elements. Once again using water as our example, if the hydrogen is the first isotope we get ordinary water. When it is the second or third isotope we get heavy water because of the deuterium or tritium forms of hydrogen.

All of the materials (polymers) used in blow molding today are compounds of basic elements and compounds processed and developed by the petrochemical industry. Basically, they are based on hydrogen and the class of material is generally referred to as hydrocarbons. Resin manufacturers develop special catalysts and processes to tailor make compounds that will suit the needs of their customers. By way of explanation, typical hydrocarbons such as polyethylene:

$$2\ CH_2=CH_2(R)\ \ -CH_2-CH_2-\ \ +\ \ -CH_2-CH_2-\ (R)\ \ -CH_2-CH_2-CH_2-CH-2-$$

ethylene monomer　　　　　　　　monomer　　　　　　　　　　　　　　dimer

can join upon application of heat and in the presence of special catalysts. The process shown above converts two (2) molecules of ethylene into two (2) activated monomers that can join to form a dimer of ethylene. Continue this process thousands of times and a multiunit (polymer) chain of polyethylene will form. The catalyst controls the way in which the polymer grows and thus its final bulk properties. At high temperatures the chains of polyethylene are long coils. When melted, they are moving very rapidly. When cooled, they lose motion and become "glassy." They are clear at this point. Then something happens, in a way not unlike an old-fashioned carpenter rule, the chains start folding back on themselves and become crystalline. The previous clarity is lost and the material now becomes cloudy or opaque. While the material is cooled to ambient temperature, the state or morphology of the polyethylene becomes psion, blow molding or injection molding, it occurs in the feed and compression sections of the screw pushing the polymer into the mold. As the polymer cools in the mold, the crystalline domains that existed in the pellets or flakes being molded reform as the molded part cools. The change in volume that occurs during recrystallization accounts for the mold shrinkage for these materials. Crystallization is influenced by the cooling time, melt temperature, and other factors. Variations in crystallinity in the part caused by these factors can introduce internal stress and warpage, often weakening the part.

12.2 Polymers

Polymers or resins (naturally produced products from trees and plants) are formed by a variety of chemical processes generally referred to as polymerization. This means that one or more types of less complex molecules are combined to form larger, much more complex molecules. Most importantly, these molecules or chains can be selectively produced to yield polymers or plastic materials of varying hardness, strength, color, weather resistance, or other properties to meet a wide range of applications. As previously noted, the monomer is the basic repeating unit in the polymer chain. It contains the elements of the final plastic. The normal form of a monomer is either a gas or liquid. It chemically unites under the right conditions with itself or other monomers to form the desired polymer.

12.2.1 Homopolymers, Copolymers, and Terpolymers

A polymer made up exclusively of a single monomer is called a homopolymer. Both polyethylene and polypropylene are often homopolymers. It was found that one can easily mix ethylene and propylene, the monomers for polyethylene and polypropylene respectively and get a completely different series of polymers. This type of polymer made up of two monomers is known as a copolymer. The properties are different from either homopolymer and thus become an important way to tailor make a desired polymer for an application. Another example to consider are monomers such as acrylonitrile, butadiene, and styrene. These three different monomers can be joined together to make three types of copolymer and two types of a more complex structure called a terpolymer. Examples of the latter are ABS plastic and nitrile rubber, both terpolymers in wide use today.

12.2.2 Thermoplastic and Thermoset Polymers

As previously noted, polymers may be classified into two major groups based on physical performance, thermoplastics and thermosets. Thermoplastic resins may be softened or melted and reshaped repeatedly by application of pressure and heat. This allows various forms and sources of reclaimed polymer from products to be reground and remolded many times. Thermoset materials, linked chemically during molding will degrade at a high temperature before molecular motion allows them to melt. This allows these materials to perform well in high temperature environments such as cooking appliance components and handles, automotive engine parts, electrical components for circuits and aerospace applications.

12.2.3 Amorphous and Crystalline

Thermoplastics can also be grouped into two structural categories, amorphous and crystalline. In an amorphous resin the molecules exist in a random state. These long chains are intertwined with each other, forming a glassy mass. In general, amorphous resins have less shrinkage when cooled. Processing amorphous plastics is generally easier because of a large melt temperature range which helps to minimize stress molded in a product. Products made from these resins can be rigid with low to moderate impact strength (for example polystyrene or methol methacrate), or they can have excellent impact and clarity (as found in polycarbonate).

In a crystalline polymer, molecules orient in a more ordered fashion. The polymer chains essentially lie side by side in an orderly fashion. Crystalline polymers are usually tougher, softer, and have high shrink rates when cooled. The melt ranges of these types of materials are usually narrow. Examples of crystalline resins are polyethylene and nylon. Polypropylenes exhibit a range crystalline and amorphous behavior depending upon stereoisomer foam or a blend of them (Fig. 12.1).

12.2.4 Fundamental Properties

The physical performance of polymers can be summarized by characterization using a series of tests. An understanding will now be developed on how materials are classified (grade) by testing and how this information is used to pick an optimal material for a particular application. Throughout this section the designer needs to ask the following three questions:

1. What are the characteristics or properties of this material?
2. What role will these characteristics play in how the parts are molded?
3. Will these characteristics make the material suitable for the application?

The designer should also keep in mind four fundamental material concepts that could be the source of problems. This will allow the performance difference or variance to be traced:

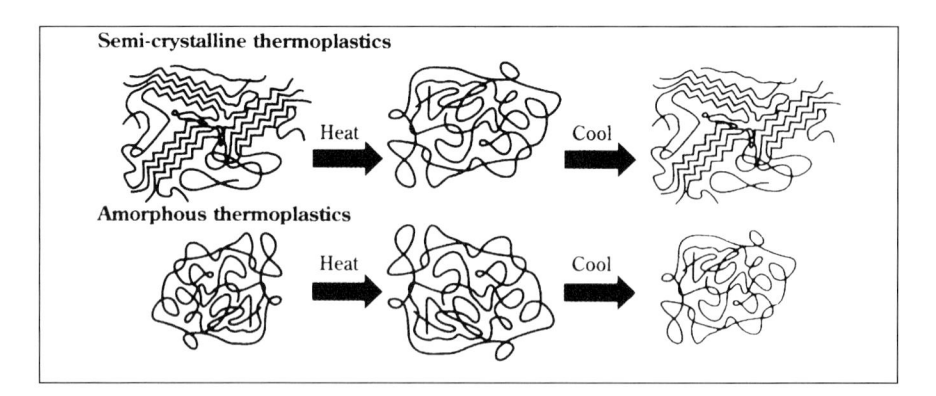

Figure 12.1 Most thermoplastics are either amorphous or semicrystalline. The crystalline regions of semicrystalline polymers, such as nylon or polypropylene, melt during plastication and reform during the cooling phase of the molding process. Amorphous polymers, such as polystyrene or acrylic do not exhibit this transition which means it does not crystallize, hence both materials have high clarity

1. Average molecular weight,
2. Morphology,
3. Chain linking, and
4. Additives, fillers, and reinforcing agents.

12.2.4.1 Average Molecular Weight

Plastic materials are chainlike molecules made up of a repeating link or "mer." The number of links that make up the chain (i.e., the chain length or molecular weight) can vary considerably from one grade to another. Most materials are available in a wide variety of molecular weights, or more correctly, average molecular weights. Unlike some other chemicals, plastic material grades do not have a fixed molecular weight, but rather have a distribution of variable chain lengths leading to an average molecular weight as indicated in Fig. 12.2.

The material grades with the higher average molecular weights tend to have better performance properties (creep resistance, chemical resistance, impact strength) due to an increase in chain (molecular) entanglement and intermolecular attraction (i.e., the attractive forces between adjacent polymer chains). Those with lower molecular weights tend to offer improved processibility or lower melt viscosity, but somewhat lower material performance. In practice, plastic engineers and designers commonly refer to melt flow rate (MFR) or melt index (MI) as measures of molecular weight.

12.2.4.2 Chain Length Linking

As stated, plastic materials are chainlike molecules made up of repeating links or mers as indicated in Fig. 12.3. Like molecular weight, the chemical structure of the repeating unit or link will influence the properties of a material. Both the degree of polymerization (chain length/molecular weight of monomer) and the repeating unit structure of a plastic material

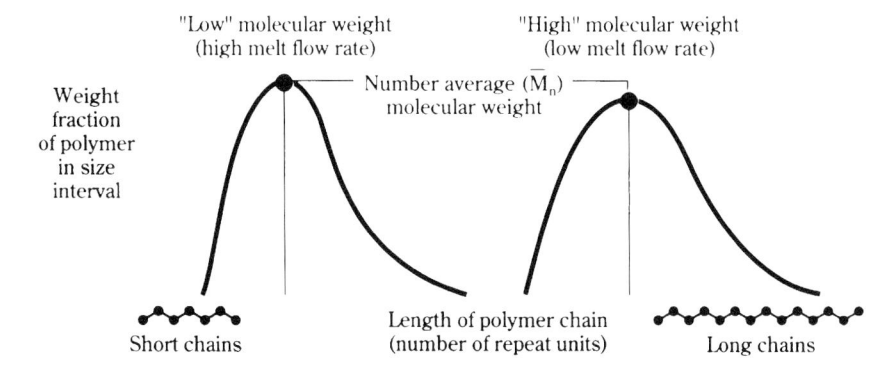

Figure 12.2 Most thermoplastics are available in a wide variety of average molecular weights (chain lengths). Those grades with lower average molecular weights (higher MFREs) offer improved flow characteristics, while higher molecular weight grades tend to offer enhanced performance characteristics

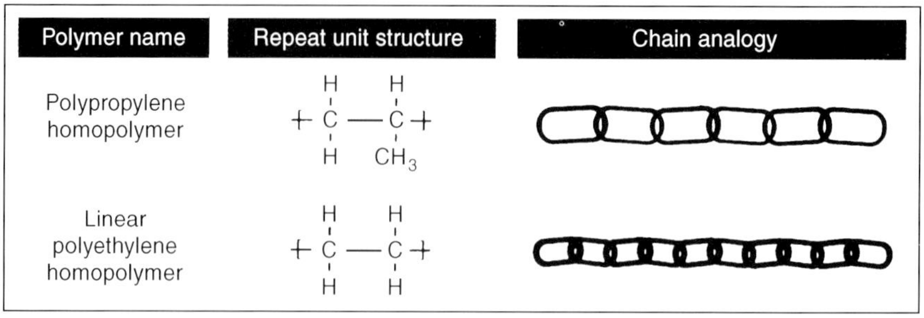

Polymer name	Repeat unit structure	Chain analogy
Polypropylene homopolymer	$+\overset{H}{\underset{H}{C}} - \overset{H}{\underset{CH_3}{C}}+$	
Linear polyethylene homopolymer	$+\overset{H}{\underset{H}{C}} - \overset{H}{\underset{H}{C}}+$	

Figure 12.3 Both the degree of polymerization (chain length/molecular weight) and the repeat unit structure of a plastic material influence the material's end use properties. Two materials with the same chain length, but different repeat unit structures, will have different properties

influence the material's end use properties. Two materials with the same average chain length, but different repeating unit structures, typically will have different properties.

12.2.4.3 Morphology

Morphology is the way that the chains arrange themselves relative to one another. The most common chain arrangements associated with thermoplastics are random (amorphous) and semicrystalline. The molecules of amorphous polymers, such as polystyrene, are in theory random coils with no particular orientation. The materials soften when heated due to a reduction in intermolecular bonding, and gradually regain their strength as they cool in the mold. However, in practice, it is likely that there is at least some degree of "frozen-in" orientation associated with the injection of hot melts into relatively cold molds.

12.2.4.4 Additives, Fillers, and Reinforcing Agents

Almost all commercial plastic materials contain added chemical fibers. Like medication, they are beneficial in some respects, but can have negative side effects as well. This is particularly true when glass fibers are used at higher concentrations. Glass fibers are commonly used to enhance the stiffness, strength, and dimensional stability of plastic materials. However, the side effects associated with the use of glass fibers include factors such as increased abrasion during processing, as well as a deterioration in the surface finish achievable. Other examples include antistates, antiblocking agents, and slip agents such as GMS (glyceryl monosterate) and euricicamide. These make part surface slippery and difficult to print on. A general understanding of these material factors will enhance the designer's ability to establish a suitable design, and avoid unanticipated processing or performance related problems. As a simple example, consider the embrittlement of a molded thermoplastic part operating in a high-temperature environment that is subject to intermittent loading (e.g., a latch mechanism). Initially, the part functions properly, but over time becomes brittle and fails under load. There are a number of possible causes of such aging related failure, including: a possible change in structure or molecular weight due to oxidation, hydrolysis, and/or ultra violet light; a change in morphology (particularly for a semicrystalline material; or a change in additive concentration (migration/evaporation of volatile such as a plasticizer or even water). While there are many similar examples, the key point is to keep each of these four plastic material fundamentals in mind when making design related decisions or when diagnosing problems.

12.3 Physical Properties

The physical properties of polymers determine selection for use in a product. Among the properties tested for in resins are average molecular weight, hardness, flexibility, tensile strength, flow, moisture content, and impact resistance. To gain a better understanding as to why a material is chosen for a specific job, the following steps are taken into account.

12.3.1 Specific Gravity

By simply dropping a plastic sample in water we can gain an idea of its type and its specific gravity. Specific gravity is defined as the ratio of the weight of a given volume of a substance to that of an equal volume of another substance, usually water. In simpler terms we are comparing the difference in weight between resins when suspended in water. If the sample material floats in water it may be polyethylene or polypropylene. These materials have a specific gravity less than water. Water is always referred to as 1.0 PE and PP are in the 0.90 to 0.96 range. If the sample material sinks in water it may be any of the materials that has specific gravity greater than water. The specific gravity of a material can be used, multiplied by the head's capacity of PE weight to calculate an accumulation heads capacity.

Example: Specific Gravity of PC = 1.20

Head capacity   × 15.88 kg (35 lbs)
Head capacity in PC = 19.05 kg (35 × 1.20 = 42 lbs).

Parts molded of one resin vary in weight when compared to the same part molded in another material by the ratio of their specific gravities.

Specific gravity listing of materials:

Phillips HMX50100 HDPE	950 9/cm^3
Allied BA50100 HDPE	950
Soltex K-50-10 HDPE	950
Phillips HGX-010 polpro	904
General Electric Cycolac ABS	1.020
General Electric Noryl PPO	1.100
General Electric Lexan PC	1.200
General Electric Prevex PPO	1.100
General Electric GTX PPO	1.100

General Electric Geloy 901.................... 1.060
Dow Magnum 941 ABS.................... 1.050

12.3.2 Melt Flow Rate (Melt Index)

The melt index of a resin helps estimate its flow behavior during processing. It is commonly called MI, will vary with the molecular weight of the material. The MI is defined as the number of grams of resin forced through a 2.095 mm (0.0825-in.) orifice (die) when subjected to 2160 g of force in 10 minutes at 190 °C (374 °F). If the MI of a resin is high, the melt flow resistance during processing is lower than when the MI is low (Fig. 12.4). MI specifications are an important quality parameter of a resin. If material is displaying poor hang strength or unusually low motor current, it could be caused by a higher than specified MI.

12.3.3 Moisture

Resins that absorb moisture are called hygroscopic. The moisture content of a resin must be considered before processing for hygroscopic materials, such as PET and nylon. High

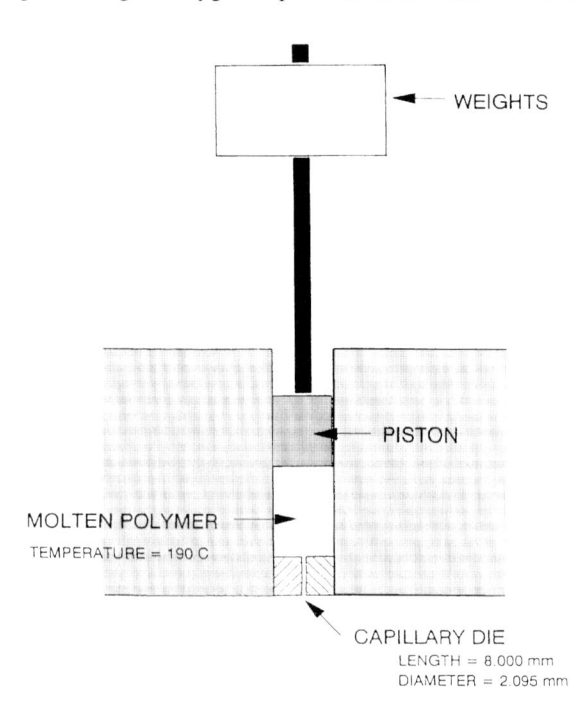

Figure 12.4 Melt flow test

moisture or "wet" resins can cause blistering, weakened melt strength, and poor part integrity. With many plastics, moisture must be kept below 0.2% by weight. This number is determined by the level of how much water is absorbed in 24 hours at room temperature, if greater than 0.1% drying is required. It is noted that PE and PP are not hygroscopic and it is necessary to make sure they are not wet due to condensation, rain, etc. A popular method for drying materials is the hot air desiccant process. Air is passed over and through a bed of moisture absorbing desiccant. This removes the water in the air and lowers the dew point. This makes air very dry and causes it to pull moisture from the resin. A hopper drier can be mounted on the machine to keep material dry after it has left the primary unit. These units utilize hot air only.

12.3.4 Hardness

Hardness is another important factor in the selection of a polymer. In general, hardness refers to scratch and abrasion resistance, and is often associated with other properties. Material can be hard and tough, hard and brittle, or hard and strong. The test for hardness is usually made by the indentation of a pin into the plastic surface. Measurements are made by "Rockwell" hardness testers or "Shore" durometers. Refer to the chart below for comparison of hardness values based on the Rockwell Hardness Scale.

Softer Materials		Harder Materials	
130	Polycarbonate	140	
120	PPO/Noryl	130	Phenolic
110		120	
100		110	
90	Polypropylene (homopolymer)	110	Polyester
80		90	
70		80	Polyvinyl chloride, acetal
60		70	
50	Polyurethane (flexible)	60	
40		50	
20	HDPE		

12.3.5 Tensile Strength and Properties

Tensile strength refers to the resistance of a plastic part to being pulled apart. The tensile strength of plastic materials ranges from 2000 to 30,000 psi (137.9 to 2068.5 bars). Tensile properties are one of the most important single indications of a plastic's strength. A tensile testing machine can reveal a polymer's "yield strength," the point where nonreturnable stretch begins. "Ultimate elongation" or the maximum length a material can stretch before breaking is also measured. Associated with tensile strength is the term "toughness." This is defined as the total energy that is needed to break the sample. The tougher the material the more difficult it is to break. High tensile materials with good elongation such as polycarbonate and ABS are used in applications where impact is important. High tensile materials such as polystyrene that have low elongation are too brittle for some applications unless rubber modified, which will tend to lower tensile strength.

Tensile strength of plastics:

High-density PE.................... 3500 psi (241.33 bars)
Polypropylene........................ 3800 psi (262.01 bars)
ABS.. 5500 psi (379.23 bars)
Polycarbonate........................ 8000 psi (551.60 bars)
Nylon.................................... 15,000 psi (1034.25 bars)

12.3.6 Creep

Creep refers to the slow dimensional change of a plastic material when it is placed under load for a long period of time. Creep resistance is an important property for a product that will carry a heavy load or must be serviced. Temperature is an important factor in creep.

12.3.7 Basic Polymer Parameters and Their Effect on Product Properties

See Fig. 12.5.

BASIC POLYMER PARAMETERS

AND THEIR INFLUENCE ON CONTAINER PROPERTIES

	DENSITY		MOLECULAR WEIGHT		MOLECULAR WEIGHT DISTRIBUTION	
	INCREASES	DECREASES	INCREASES	DECREASES	BROADENS	NARROWS
ENVIRONMENTAL STRESS CRACK RESISTANCE (ESCR)	↓	↑	↑	↓	↑	↓
IMPACT STRENGTH	↓	↑	↑	↓	↓	↑
STIFFNESS	↑	↓	—	—	—	—
HARDNESS	↑	↓	—	—	—	—
TENSILE STRENGTH	↑	↓	—	—	—	—
PERMEATION	↓	↑	—	—	—	—
WARPAGE	↓	↑	—	—	↑	↓
ABRASION RESISTANCE	—	—	↑	↓	—	—
FLOW PROCESSABILITY	—	—	↓	↑	↑	↓
MELT VISCOSITY	—	—	↑	↓	↓	↑
COPOLYMER CONTENT	↓	↑	—	—	—	—

Figure 12.5 Basic polymer parameters (Courtesy of Hoechst)

12.4 Characteristics for Blow Molding

12.4.1 High-Density Polyethylene

- High impact strength,
- low temperature toughness,
- excellent resistance to chemicals,
- good electrical insulating properties, and
- poor ultraviolet resistance.

No drying required
Melt temperature range: 190 °C to 235 °C (370 °F to 450 °F)
Typical barrel temperature profile:

Feed:	190 °C to 200 °C (370 °F to 390 °F)
Transition:	195 °C to 205 °C (380 °F to 400 °F)
Metering:	195 °C to 205 °C (380 °F to 400 °F)
Head:	195 °C to 223 °C (380 °F to 430 °F)

Regrind ratio: Up to 100% may be used for two generations with minimal effect on most physical properties
Pellet type: Virgin has a cylindrical or barrel shape
Die swell: 2 or 3 : 1 depending on die design and pushout rates
Blowup ratio: 3 : 1 safely, possibly 4 : 1 depending on part design
Hang strength (a function of drop time and size): Very good when at proper process temperatures
Shrinkage: 0.15 to 0.4 mm/m (0.015 to 0.040 in/in or 1.5 to 4.0%)
Material identification: Material in natural state is milky white. Burns easily with the smell of candle wax. Has a melting point of 130 °C to 135 °C (266 °F to 275 °F). It can be cut easily with a knife or scratched with a fingernail. Temperature consideration: Try to hold melt temperature at the lower end of the range. Temperatures that fall below this range can result in nonmelt and/or a rough finish. Temperatures above this range may result in high gloss streaking.
Mold cooling: Requires temperatures of 7.2 °C to 29.4 °C (45 °F to 85 °F) to provide optimum cycle.
Cycle times: Often restricted by the shrinkage of the part.
Note: See Fig. 12.6 for properties that make the difference (focusing on high molecular weight).
High density polyethylene, for large part blow molding applications.

12.4.2 Acrylonitrile–Butadiene–Styrene

- ABS is a hard, tough material;
- has good impact resistance;
- has good electrical insulation properties; and
- has versatile additive, filler, and reinforcing agents acceptance

192 Polymers and Plastic Materials

a: Environmental stress crack resistance

b: Impact strength

c: Flexural modulus

d: Chemical resistance

e: Abrasion resistance

f: Tensile strength

g: Ultra violet protection

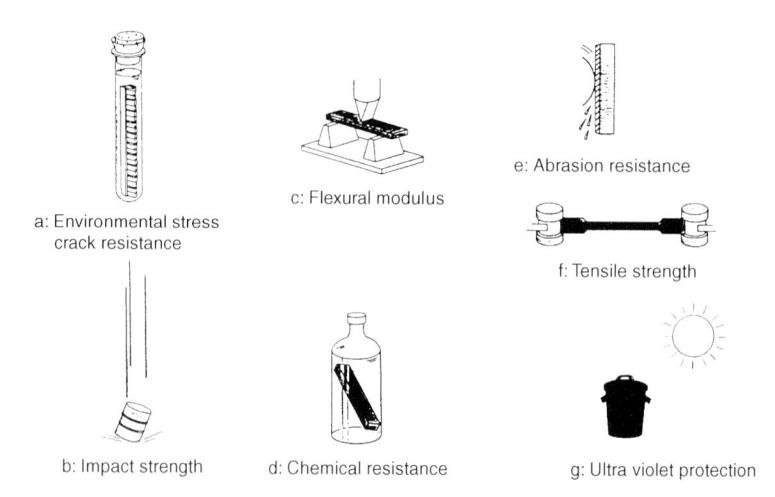

Figure 12.6 High molecular weight high density polyethylenes, designed for an optimum balance of density, molecular weight distribution, demonstrate maximum property advantages for large part blow molding (LPBM) applications. a. Environmental stress crack resistance, many large part blow molding applications require an HDPE which is very durable and highly resistant to cracking in the presence of certain chemicals, such as detergents, a HDPE with good environmental stress crack resistance (ESCR). HMW-HDPE resins, because of their high molecular weight, offer excellent ESCR for such applications as: shipping containers for chemicals, spray containers for herbicides, and pesticides refuse containers. b. Impact strength polyethylene exhibits very good resistance to failure due to impact, and the high molecular weight of HMW-HDPE resins enhances this property, especially at low temperatures. Containers made with HMW-HDPE can withstand very abusive treatment, even at temperatures below freezing point. c. Flexural modulus measures stiffness, an important property for a container during handling where it must maintain its shape without deflecting and distorting under the weight of its contents. In most cases, stiffness is obtained by increasing density and sacrificing toughness, HMW-HDPE resins deliver stiffness and toughness by increasing both density and molecular weight. d. Chemical resistance. The high molecular weight of HMW-HDPE makes it extremely resistant to the effects of a wide range of chemicals: most acids, alkalis, farm chemicals, industrial cleaning compounds, and cosmetic preparations. Shipping containers for hazardous chemicals, herbicides and pesticides require good chemical resistance and HMW-HDPE resins perform well. Substances often encountered in refuse applications can be more than adequately handled by highly resistant HDPE. Specific data regarding the chemical resistance of HMW-HDPE is available on request. e. Abrasion resistance helps prevent scuffing and gouging of refuse containers and drums during handling. Also determined by molecular weight, abrasion resistance is enhanced by the high to extra high molecular weight of HMW-HDPE resins. f. Tensile strength, like stiffness, is a function of density and determines how thick a container wall must be in order to hold a given pressure or load. Hoechst Celanese selects the density of each HMW-HDPE resin to allow the manufacture of thinner-walled containers without sacrificing the physical properties of a specific application. g. Ultraviolet protection containers manufactured for use out-of-doors must be protected from the degradation effects of sunlight by the addition of UV stabilizers (Courtesy of Hoechst A.G.)

Drying required: ABS is hygroscopic, that is, it will absorb moisture from the atmosphere. A dryer is recommended. Drying should be a minimum of 4 hours at 76.7 °C to 82.2 °C (170 °F to 180 °F). Maximum drying time is 12 hours.
Melt temperature range: 198 °C to 227 °C (390 °F to 440 °F)
Temperature profile:

　Feed: 187 °C to 193 °C (370 °F to 380 °F)
　Transition: 193 °C to 221 °C (380 °F to 430 °F)
　Metering: 193 °C to 221 °C (380 °F to 430 °F)
　Head: 193 °C to 221 °C (380 °F to 430 °F)

Regrind ratio: Up to 100% can be used but consideration must be taken that physical properties of the part are not affected. Regrind may need to be dried if not stored before reuse.
Pellet type: Virgin pellets are cylindrical in shape.
Die swell: 1.5:1 dependent on extrusion rate and head tool design.
Blow up rate: 1.5 or 2:1 dependent on part design.
Hang strength: Good compared to most engineered resins. Dependent upon melt temperature, parison weight distribution, and additive package.
Shrinkage: 0.127 mm to 0.203 mm (0.005 in. to 0.008 in.)
Material identification: Material will sink in water. Has a sweet styrene smell when dried.
Residence time: At high temperature fumes can be a problem. It remains stable for only a short period of time. If an extended delay occurs, then lower heats to 121 °C (250 °F) degradation can be reduced if shot weight is at least 80% of head capacity.
Temperature consideration: Mold temperatures may range from 23.9 °C to 29.4 °C (75 °F to 185 °F). The higher the temperature the better the texture reproduction and parting line weld strength. Part design may require that mold halves be maintained at different temperatures, because of wall thickness variation due to part configuration.

12.4.3 Polycarbonate

Polycarbonate (PC)

- Has excellent resistance to heat,
- is a hard, tough material,
- has good impact resistance, and
- has excellent transparency.

Drying required: PC is hygroscopic and will readily pull moisture from the atmosphere.
Drying temperatures: 93.4 °C to 104 °C (200 °F to 220 °F) for a minimum of 6 hours. Maximum drying times are 18 hours.
Melt temperature range: 254 °C to 271 °C (490 °F to 520 °F)
Barrel temperature profile:

　Feed:　　　　249 °C to 260 °C (480 °F to 500 °F)
　Transition:　254 °C to 271 °C (490 °F to 520 °F)

194 Polymers and Plastic Materials

Metering: 254 °C to 271 °C (490 °F to 520 °F)
Head: 254 °C to 271 °C (490 °F to 520 °F)

Regrind ratio: Avoid using ratios above 50%. Regrind has a negative effect on hang strength after generation. Regrind material must be dried before use.
Pellet type: Virgin material is cylinder shape.
Die swell: 0.8 to 1 : 1 This resin has no die swell. Will require larger head tooling than other resin.
Blowup rate: Typically 1 : 1 part design is an important factor here. Quick setup rate of this material also plays a strong role.
Hang strength: Poor stiffness at melt temperature. Parison support (possible mechanical) is critical. Top pinch bar may be required.
Shrinkage: 0.203 mm to 2.23 mm (0.008 in. to 0.088 in./in.)
Material identification: Pellets will sink in water. Material softens at 148.9 °C (300 °F) and melts at 221 °C (430 °F)
Residence time: Material will degrade quickly if flow is stopped. Keep material moving.
Degradation will occur at some point even if melt flow has not been stopped. This can be seen in yellow and brown streaking. Purging with ABS may increase degradation.
Temperature considerations: Suggested mold temperatures are 35 °C to 99 °C (95 °F to 210 °F). High mold temperatures ensure texture reproduction and improve dimensional stability of the part. Pinch edge weld also is improved.
Startup: Start up temperatures should be 11 °C (20 °F) higher than normal operating temperatures, so as to reduce high screw load caused by cold material feed. Temperatures can then be lowered when a proper parison is produced.
Shutdown: Run machine completely dry. Purge with a high molecular weight resin. Purging with ABS may increase degradation. As temperatures are brought down the cooling PC will pull contamination off the interior walls of the extruder and head. Head disassembly may be necessary after long runs in PC.
Cycle time: Because of the quick setup properties of PC cycle times are faster than for many other materials.

12.4.4 Polypropylene

Polypropylene (PP) has

- Good impact strength (poor at cold temperatures),
- good chemical resistance,
- high abrasion resistance,
- high melt strength, and

Drying required: None
Melt temperature range: 190 °C to 232 °C (375 °F to 450 °F)
Barrel temperature profile:

Feed:	187 °C to 199 °C (370 °F to 390 °F)
Transition:	199 °C to 226 °C (390 °F to 440 °F)
Metering:	199 °C to 226 °C (390 °F to 440 °F)
Head:	199 °C to 226 °C (390 °F to 440 °F)

Regrind ratio: Up to 100% regrind can be used for two generations. In polypropylene that is loaded with white pigment, yellow streaking may appear when regrind is used. Lowering temperatures by 5.55 °C to 11.1 °C (10 °F to 20 °F) may help.
Pellet type: Virgin pellets are cylindrical shaped.
Die swell: 2 or 3 : 1 depending on die design and extrusion rate.
Blowup ratio: A maximum of 3 : 1 for best molding performance. Mold design may decrease this ratio.
Hang strength: Very good at proper melt temperatures.
Shrinkage: 0.3 mm to 0.33 mm (0.012 to 0.013 in./in.)
Material identification: Virgin material is "milky" in color. Pellets will float in water. Pellets have hard, dry feel. Has a candle like smell when burned. It will begin to melt at 170 °C (338 °F)
Residence time: Similar to PE but will degrade more readily. If melt flow must be stopped be sure to fill and purge head every 20 minutes.
Temperature considerations: Mold temperatures can be in the 5.5 °C to 23.8 °C (50 °F to 75 °F) area to provide fastest cycle. Mold texture is best reproduced at these temperatures.
Cycle times: Like most materials with high shrink rates, parts made of PP are often cycle restricted. Use lowest mold temperatures possible that will still provide desired surface finish.

12.4.5 Polyphenylene Oxide

Polyphenylene oxide

- Has good flame retardancy,
- has good chemical resistance,
- has good impact resistance, and
- retains mechanical properties in high heat environments.

Drying required: Dry for a minimum of 4 hours at 82.2 °C (180 °F). Maximum drying time is 18 hours. Regrind will need to be dried if not used within an hour.
Melt temperature range: 210 °C to 232.2 °C (400 °F to 440 °F)
Barrel temperature profile:

Feed:	199 °C to 210 °C (390 °F to 410 °F)
Transition:	204 °C to 221 °C (400 °F to 430 °F)
Metering:	204 °C to 221 °C (400 °F to 430 °F)
Head:	204 °C to 221 °C (400 °F to 430 °F)

Regrind ratio: Up to 100% regrind may be used in some applications. If possible, remain below 50 for best melt strength.
Pellet type: Virgin pellets are cylindrical in shape. Pellets often come precolored.
Die swell: 0.9 to 1 : 1. Die swell marginally better than for polycarbonate. May require larger head tooling than most resins. Diverging tooling will be required in most cases.
Blowup ratio: 1 or 1.2 : 1 depending on part design.
Hang strength: Marginally better than polycarbonate. Staying on the low end of the melt range will provide best stiffness. Parison support is required in most cases.
Shrinkage: 0.06 mm to 0.09 mm/cm (0.006 in. to 0.009 in./in.)
Material identification: Pellets will sink in water. Material has a strong styrene smell when processed. Smoking may be a problem when melt temperatures are too high.
Residence time: Material will begin to degrade in 12 hours if melt flow is stopped. Purge every half-hour to avoid degradation.
Temperature considerations: Mold temperatures can range from 22.2 °C to 95 °C (70 °F to 185 °F). Higher mold temperatures provide better reproduction and weld strength.
Cycle times: This material sets up quickly and provides for fast cycle times. Wall thickness above 2.286 mm (0.090 in.) will require cycles.

12.4.6 Polyethylene Terephthalate

Used in injection blow molding, produces clear amorphous preforms. Since there is less orientation the impact strength is reduced and used extensively for bottles up to 340.5 g (12 oz.) and wide mouth containers. By subjecting an injection preform part to mechanical deformation below melting point, it produces a biaxially orientated, tough article that is used for deep drawn containers. A major factor in the market for carbonated bottles is that the properties are balanced.

12.5 Coloring Plastic Materials

Thermoplastic materials, in general, can be molded in a wide range of colors. The color can be provided in a precolored base plastic or by adding solid or liquid color concentrates at the machine just prior to plastication.

Color concentrates are high pigment dispersions of colorants in carrier resins. The concentrate supplier matches the desired color by using a blend. The supplier then compounds a concentration formula, typically 20 to 60%, in a carrier resin. Additives such as antioxidants, stabilizers, and antiblocking agents are often coblended at this point. A recommendation is then made as to the amount of concentrate required to blend with the base resin to obtain the desired color. This is referred to as the letdown ratio. Example: if 454 kg of resin is mixed with 454 g of concentrate the ratio or let-down is 100 : 1. Generally, the lower the ratio (25 : 1, 30 : 1), the easier it is to disperse the color accurately. Factors such as screw *L/D* ratio and screw speed may also require higher or lower percentages of pigment than originally suggested. The reason for this is that the pigments need to be mixed by the screw, and thus the less time the resin must spend in the extruder. The carrier resin is preferably the same as the letdown resin for good compatibility. The melt flow of the carrier resin should generally be higher so that it will mix readily and uniformly throughout the letdown resin.

The important quality check for color concentrates is the color match. The molded part must match when compared to the color chip and color references. It is important that the type of light used for matching is specified. Ultraviolet, fluorescent, and sunlight are common choices. Colors may match under one source of light but not another. This is known as metamerism. Color concentrates are heat sensitive and may reduce available residence times. Generally, reds are the most heat stable while yellows and oranges work in the moderate range. Darker colors require lower processing temperatures. Color concentrates that have a hygroscopic base resin may need to be blended before drying.

12.6 Regrind

Extrusion blow molding by its nature will generate a certain amount of flash. In some parts flash can be up to 100% of the total part weight. It is an economic must that this flash be recovered as regrind. For regrind to be used it must be kept clean. Foreign material can harm surface appearance and degrade the properties of the part or resin. Foreign material may also hang up in the head and cause additional problems. All material should be covered and all material handling equipment (grinders, boxes, loader) be kept clean. The amount of regrind used in a given product is determined by several factors. For the properties that make a difference see Figs. 12.6a to g.

12.6.1 Regrind Specifications

Many times testing of physical properties is required to determine a maximum or optimal amount of regrind in the finished product. Reprocessing material tends to reduce these values.

12.6.2 Process Performance

For some materials, regrind levels above 50% can have a negative effect on hang strength and die swell. Poor parison length consistency is a result of these problems.

12.6.3 Physical Properties

Some resins will lose important physical characteristics at high regrind levels. Most resin suppliers recommend that a regrind have no more than three heat histories. Parts that must withstand impact or repeated stress should be produced with precisely controlled levels of regrind. Regrind (or post-consumer recycled material, PCR) is applied in the middle layer in coextruded parts. Reduce flash by keeping the mold as close to the head as possible and by using a properly sized head tool for the job. Regrind with a high degree of dust or "fines" may require feed zone temperatures to be reduced by 5.55 °C to 11.1 °C (10 °F to 20 °F). This will help prevent premature melting of the plastic. Regrind percentages will change the percentage of color master batch that is added, the processor's ability to match a specified color. This is because many colors are heat sensitive and will discolor after being reprocessed. Regrind material will convey differently than virgin pellets, because of the irregular shape of the reground pellets. Materials that are "rubbery" in feel will tend to stick together.

12.7 Post-Consumer and Industrial Recycled Materials

Most of the factors to be considered in processing are the same for recycled HDPE and virgin HDPE. Most other materials are not blow molded in volumes that are finding their way into the recycling stream. HDPE in a variety of grades, especially for blow molding, is the largest volume plastic available. Primary consideration in handling recycled material is that chip or flake will flow differently than pellets. This will cause a problem in the initial feed area due to more chance of bridging at the extruder inlet. The amount of fines in flakes is increased this situation. Determination of the amount of fines is often important in order to evaluate the potential for this type of problem. The problem of optimizing concentration level of the recycled HDPE to the virgin HDPE in a given application centers on the long-term as well as the short-term property requirements. Exposure to accelerated or natural ultraviolet (UV) radiation of test bars of the material is necessary to establish the long-term toughness of the material and compare it to the initial value for the HDPE. This will ensure the selected material ratio will meet the short-term and long-term requirements for the product. The same procedure may be duplicated for other recycled plastic materials. The blending of virgin and recycled material together, and the determination of what percentages of blend will meet the specifications as they are normally applied to virgin materials. The feed streams for HDPE and other plastics will give different characteristics. Staying with HDPE, milk jugs, for example, are near the low end of the flow spectrum with fractional melt indices. In other examples, such as detergent bottles and mixed colors, melt indices of fractional to 5 or so are common. This is good for blow molding, but bad for injection molding where high flow (melt index of 20 to 30) is necessary. Films (trash bags, pallet wrap, etc.) also are poor for injection molding but good for extrusion or blown film applications. The point is, testing must be done to characterize a given stream of recycled material. This will ensure that the proper recycled material gets into the proper production part or product.

Acknowledgment

The author wishes to thank Ron Walling, Advance Materials Center, for reviewing this chapter and making suggestions for its improvement.

12.8 References

1. Walling, R., SPE Large Part and Industrial Blow Mold Seminars (Materials Session), Advanced Materials Center, Inc., Ottawa, IL
2. Mallory, R.A., Design Watch, Remember the Four Material Fundamentals, Plastic World. (Nov. 1995). (Author's note: Although this work is directed toward injection molding the same basic principles apply here.)
3. Hartig Training Manual. Davis-Standard Corporation, Edison, NJ

13
Cost Estimating

13.1 Introduction

It is desirable for the designer/engineer to have a general understanding of how the cost of a blow molded part is calculated, as well as the elements that are considered.

The marketing group will provide the sales forecast.

The process engineer will be responsible for the machine selection, cycle estimates, material selection, weight and usage calculations, labor, component selection, and bill of materials.

Purchasing is responsible for buying the materials and component cost.

The cost accountant is responsible for the cost sheet format, variable and fixed cost allocations and the final cost sheet calculations (see Chapter 2).

This chapter is based on the experience of the author, and is intended as a guide only, not an introduction on cost accounting. Each company may approach cost accounting slightly differently. This information may be sought from your cost accountant in order to gain further insight into your company's methods.

13.2 Typical Cost Sheet

A typical cost estimating sheet may consist of materials (resins), machine cost, machine operator (labor cost), purchased parts (other than resins), finishing labor (other than the machines), S.G. & A. (Sales, General Operating, and Administrative expenses). Figure 13.1 shows a typical cost sheet which breaks down as follows:

A. Materials

Weight calculations are made for each element of material used. There are allowances for loss due to storage, handling, and manufacturing (usually based on experience records)

There is an allocation for overhead representing storage, handling, and manufacturing (based on company annual budget and allocated as a percentage of total polymer weight processed or another acceptable method).

B. Machine/Molds
Type of clamping force

A: MATERIALS

TOTAL WEIGHT INCLUDED COLORATION AND ANTI-UV (Kg)
PLASTIC COST/Kg
COLORATION
    % (total weight)
    Coloration cost/Kg

Anti-UV
    % (total weight)
    Anti-Uv Cost/Kg
Material losses (% total weight for storage, manutentions...)
Overhead (purchasing, storage, manutention...)

TOTAL MATERIALS

B: MACHINE MOLDS

Type of (clamping force) (T) _____

Operating hours
    22 h/day - 5 days/week - 49 weeks/year - 0.8 - .095 (h)

Variable cost (Kf)
    Energy
    Utilities
        Total variable cost

Fixed cost
    Maintenance (pc)
    Taxes (local or regional)
    Insurances
    Building (+)
    Supervisors + process engineer, plant + manager maintenance + technicians
    Scheduling
    Quality Control
    Plant methods
    Purchasing
    Depreciation:

        Machine
        Molds
        Tools
        Robots

    Total Fixed Cost

Machine Rate Varible (F/H)
Machine Rate Fixed (F/H)
TOTAL RATE MACHINE/MOLDS

C: LABOR

Number
Cost/hour

D: GENERAL DATAS

Cycle (s)
Cavities
Shots/hour
Yield/hour
Finishing Rate/Hr.

E: COST

A: Materials
B: Machine
C: Labor
D: Varible Cost
    Fixed Cost
E: Finishing - Operators

FACTORY -TOTAL COSTS

S.G. & A.

TOTAL

GENERAL TOTAL

Figure 13.1 Typical cost sheet

The type of machine indicated in blow molding is usually designated by machine clamp force or shot size and/or extruder diameter or rate of material output per hour. 2)
Operating Hours 4)

Number of productive hours for each day, operating days per week, operating weeks per year (allowing for vacation, shut down, and holidays)

Percentage of actual operating hours (allowing for maintenance and repair down time, also mold changes and setup)

Production efficiency (time deductions allocated for scrap parts)

Variable cost

Energy (electrical usage)

Utilities (water, air, etc.) (Allocated from yearly budget based proportionally on machine size, horse power, etc.)

Fixed cost

Maintenance (replacement parts)

Taxes (including local and regional)

Insurance

Building

Supervisors

Scheduling

Quality control

Plant methods

Purchasing 1)

Depreciation 3)

Notes on fixed cost:

1. Purchasing may be allocated to S.G. & A.
2. Cost may be allocated to each work center based on specific identification, relative replacement, or some other factor.
3. Depreciation is the initial cost of machine and installation divided by estimated life—the number of years depending on the company's accounting policy.
4. A volume assumption must be made to derive the machine hour rate. The "normal" capacity of the work center could be used. Normal capacity is maximum capacity reduced by normal down time, holidays, vacation, shutdowns, etc.
5. A total machine rate/hour for variable and fixed costs is calculated for the budget year.
6. Note: 1. Mold, tools, fixtures, and support equipment (i.e., robots) may be determined as part of fixed cost or used in Return On Investment (ROI) calculation for the product. Another accounting term, Return On Assets (ROA), is sometimes employed.
7. If the company is the original part manufacturer, the tools, molds and special equipment would normally be quoted separately and paid for by the customer.

C. Labor

Operator cost is determined by wages per hour and benefits (hospitalization, social security, etc.).

D. General Data

The engineering group, in addition to determining the type and weight of material, provides data on operators, cycle time, number of cavities, output on shots, and yield per hour.

If there is finishing, assembly of several parts, or decorating, calculations for the number of operators, their pay rate per hour, and cost of purchased materials would be calculated, and included in the total product cost.

E. Total Cost—summary including SG & A by accountant

13.3 Cost Conclusion

Costing systems, weather absorption, direct costing, or some other method, represent the allocation of the elements and activities of a manufacturing organization. Methods may vary considerably from company to company depending on the size, management style, and accounting techniques. The foregoing has been presented to allow the designer/engineer to formulate an understanding of the cost involved when developing a blow molded part design.

13.4 Cost Estimating Calculations

In previous chapters the elements that make up the cycle time for a blow molding product have been discussed. Cycle time course is going to vary with the type of product, size, weight, machine, etc. The following is a summary of key production factors common to all products.

13.4.1 Part Requirements

1. Size and weight of part calculations.
2. The size and weight of part will determine required press dimensions for the mold to be mounted. Size of extruder and the throughput rate of the machine.
3. Quantity of parts to be made and required delivering schedule will determine the number of cavities, which in turn will affect (B) size of press and machine.

13.4.2 Cooling of the Part

The greatest amount of time in the molding cycle is spent cooling the part in the mold. The material type, weight, and wall thickness are the major considerations, particularly wall thickness. At this point, mold construction, part material, cooling channels and zones, and part and flash thickness, which should be minimized particularly in the "pinch off" area, must be determined.

13.4.3 Throughput of the Machine

Total cycle time will include the extrusion time needed to produce shot weight, the parison drop time, opening and closing of the press, and mold close time (blowing and cooling time). *Note:* If using a shuttle press or wheel machine the blowing and cooling time normally will be the cycle time; in other words, the drop time is not part of the overall cycle.

13.4.4 Post-Molding Operations

Auxiliary operations now need to be considered. The part has to be removed from the press and deflashed, cut, machined, or drilled. The time to complete these operations should be estimated to determine if they can be performed at rates equivalent to the estimated molding cycles previously calculated. This will determine if one or more operator stations will be necessary.

13.4.5 Setup and Purging of Material from Previous Product Run

The machine time for this element is usually included in the overall machine utilization calculation for the yearly operation. The setup labor for each production order usually has not been allocated; thus, estimation for the labor for this element needs to be determined, keeping in mind the purging of material (depending on color change—for example, changing from black color to beige) may take several hours to remove streaks.

This estimate becomes significant in the overall cost of the order only if quantities are small, say, less than a 5-day production run. It is however, important to a custom molder to consider.

Acknowledgments

Thanks to Certified Public Accountants Andy and Donna Lee, Greensboro, NC and Steve Andrews, ZARN, Inc., Reidsville, NC for their review and input for this chapter.

Index

Additives 89, 186–187
Aerosol containers 51
Alloys, growth of market in 7–8
Auto–blow twin head machines 4
Automobiles, application of blow molding to 70–71, 72
Average molecular weight 185

Barrier screw 105
Battenfield–Hartig. *See* Davis Standard
Beryllium–copper molds 151
BFT collapsible pallet container system 146
Blends, growth of market in 7–8
Blowing air, injection of 158
Blow molding 1
– advantages of 8–9
– history of 2
– market, growth of 6–7
– overview of 74
– products of 5
– simulations 77–78, 79
– *See also* Coextrusion blow molding; Extrusion blow molding; Injection blow molding; Polymers, blow molding of; Stretch blow molding; Three–dimensional blow molding
Blow molding, process of 1, 3, 59
– components of 6
  – molded–in insert 63–64
  – containers, configuration of 66–67
    – flat sides 67
    – lip 67–68
    – nesting and stacking 68–70
    – separating 69–70
  – engineered materials 62–63
  – foam–filled 60, 61, 62–63
  – hollow parts 59
  – interlocking systems 64, 65
  – kinetic energy 62–63
  – multiple/combination cavities 66
  – resin/fiberglass layup 60
  – snap fits 65, 66

Blow pins/needles 119, 120
– and cooling 153–154
– design of 159–161
Blow ratio 30–35
Borealis 143
Borden Dairy 4
Bottles, design for
– concepts for 44–46
  – cross–sections 47–48
  – neck, threads, and openings 48, 49, 51
  – ribs 46–47
– parison, assumptions about 44
– shapes, basic 43–44
– volume capacity of 48–49, 57
  – adjustment of 55
  – conditions of production, effects of 55–56
  – correction of 54–55
  – Dow Chemical Operators Guide 56, 58
  – measurements of 51–52
  – standards for 52–53
– *See also* Containers; Molds
Bottom plug insert 168–169
Brigham Young University. *See* Computer Integrated Manufacturing (CIM) system
Bumpers, engineered blow molded (EBM), 63
– foam–filled 63

CAD/CAM. *See* Computer aided design and computer aided engineering
Cast aluminum molds 151
Cavity oversquare 47–48
Chain length linking 185–186
Chamfers 36, 37
Closures, bottles 48, 51, 58
CIM system. *See* Computer Integrated Manufacturing (CIM) system
Clamping ring 114
C–MOLD 74–75, 84
CNC. *See* Computer numerical control machine
Coextrusion blow molding 113, 127–133

– advantages of   *127–128*
– continuous and intermittent   *129–132, 135*
– methods of   *132–133*
   – shuttling clamp   *132–133*
   – parison transfer system   *132–133*
– structures   *129, 130*
– *See also* Blow molding; Blow molding, process of
Compression molding   *59*
Computer aided design and computer aided engineering
– advantages of   *173–174*
– mold making organization, application to   *176–180*
   – engineering   *176–179*
   – manufacturing   *179180*
– systems and methods   *174–176*
   – minicomputers   *175*
   – networks   *175–176*
   – personal computers   *174–175*
Computer Integrated Manufacturing (CIM) system   *13–14, 18–23*
– enterprise wheel, elements of   *14–16*
Computer numerical control (CNC) machine   *173, 174, 175*
– programs   *179*
Computer software
– fluid flow finite element, simulation of   *76, 77–79*
   – modeling   *76–77*
   – prediction capabilities   *78–79*
– parison, simulation of   *77*
   – area/stretch ratio   *76*
   – molded   *73*
   – extrusion blow molded   *73*
   – thickness, reduction of   *74–75*
– Polymer Inflation and Thinning Analysis (PITA), *80–83*
   – primitive shapes   *80*
   – wall thickness   *80–83*
Concurrent engineering   *19*
Containers
– configuration of   *66–67*
   – warpage   *66–6 7*
– measurement of capacity of   *51–52*
– *See also* Bottles, design for
Continuous extrusion. *See* Extrusion blow molding
Cooling. *See* Molds

Copolymers   *183*. *See also* Polymers
Core rod assembly   *166, 167*
Corona discharge   *86*
Cost estimates   *201, 203*
– calculations   *203–204*
   – material, setup and purging of   *204–205*
   – part, cooling of   *204*
   – part requirements   *204*
   – post-molding operations   *204*
   – throughput   *204*
– cost sheet   *201–203*
C–PITA   *84*
CPM   *17*
Creep   *190*
Critical Path Method. *See* CPM Crushing   *63*
Cutter groove   *70, 201*
Cycle times   *102, 201–204*
Cylindrical projections   *80–81*

Davis Standard   *4*
Decals. *See* Decoration of products
Decoration of products
– decals   *99*
   – heat transfers, advantages of   *97*
– hot stamping   *97, 98, 99*
   – foils   *97, 98, 99*
– in mold labeling (IML), *100, 101*
   – aesthetics   *102*
   – cycle times   *102*
   – equipment for   *100*
   – label molds   *101*
   – process   *100, 101*
– labels   *91 , 92*
   – peel–off   *91, 92*
   – pressure–sensitive   *91, 92*
– pad printing   *95, 96*
– screen printing   *92–93, 94*
– spray painting   *89, 90*
   – etching   *90*
   – masking   *89–90*
   – sanding   *90*
   – vapor degreasing   *90*
– surface treatment   *85–89*
   – additives   *89*
   – corona discharge   *86–87*
   – etching   *88–89*
   – methods   *86*
   – flame treatment   *86, 87, 88*
   – solvents   *88*

– water-based chemicals, washing with 88
Design, parameters of 8–10
Design, process of. See Blow ratio; Computer Integrated Manufacturing (CIM) system; Flanges and tack-offs; Product Design and Development Management System (PD2MS); Radii, guidelines for; Ribs and gussets
Die 114, 115
– ovalization of 117, 118
– sets 169
Dolls, bodies of 4
Dome systems 161–162
Dow Chemical Operators Guide 56, 58

Empire Plastic 4
Engineered materials 62–63
Enterprise wheel 14–16
Etching 88–89, 90
Extruders 103–104
– screw 104–105
Extrusion blow molding 2, 3, 103
– coextrusion 113
– continuous extrusion 108–110
  – rising mold 109
  – rotary wheel 109–110
  – shuttle system 108–109
– head tooling 114–116
  – converging 114
  – divering 115
  – types, selection of 116
– intermittent extrusion 110–112
  – accumulator 111–112
  – ram 111
  – reciprocating screw 110
– mold, checklist for ordering 171, 172
– part weight and wall thickness, adjustment of 116–118
  – blow pins/needles 119, 120
  – die ovalization 117, 118
  – parison programming 117
– process of 106–107
– regrinding 197198
– See also Blow molding; Blow molding, process of; Coextrusion blow molding; Extruders

Fillers 186–187
Fischer blow molding machine 4
Flame treatment 86, 87, 88

Flanges and tack–offs 37–38, 61
– attachments to 39
– threaded parts 38, 40
Fluid flow finite element, simulation of. See Computer software
Foam filling 60, 61, 62–63
– technology of 144–145

Geometric factors 73
GE Plastic Study of Engineered Blow Molded Plastic 71, 84
Glass blowing 2
Gussets. See Ribs and gussets

Hard–soft–hard and soft–hard–soft technology 141–142
Hartig. See Davis Standard
Head tooling 114–116
Hoechst A.B. 156
Hollow structures 9, 27–28, 59
Homopolymers, 183. See also Polymers
Hot stamping. See Decoration of products

ICI 2
Ideal Toy Company 4
Injection blow molding 6, 7, 9, 10, 59
– advantages of 123
– disadvantages of 123
– machine 122–123, 125
  – three- and four-station machines 122, 123
See also Blow molding; Blow molding, process of
Injection blow molding, process of 120–121, 164–170
– design, rules for 166–167, 168
  – bottom plug insert 168–169
  – core rod assembly 166, 167
  – die sets 169
  – neck ring insert 165–166, 168
  – parison (preform mold), 164–165, 167
  – tooling summary 170
  – vents 167–168
– methods of 120, 121
– ordering, checklist for 172
In mold labeling. See Decoration of products
  Interchangeable inserts 55
Interlocking systems 64, 65
Intermittent extrusion. See Extrusion blow molding

Johnson Controls *4*

Kennedy Tool *176–178, 180*
Kinetic energy, management concepts *62–63*
Kitchen mincer *103*

Labels. *See* Decoration of products Liquid flow theory *73*

Mandrel *114, 115*
Marcus, Paul *4*
Masking *89–90*
Melt flow rate *188*
Melt index *188*
Melting *1*
Metal stamping *6*
Mills, Elmer *2*
Minicomputers *175*
Molded–in insert *63–64*
Mold making organization, structure of *176–180*
Molds
– blowing air, injection of *158*
– bottles *159–163*
  – dome systems *161–162*
  – neck ring and blow pins, design of *159–161*
  – rotating molds *163*
  – sliding-bottom *163*
– characteristics of halves *149, 150*
– cooling *151–154*
  – blowing pin *153–154*
  – cooling lines *152, 153*
  – heat transfer rate *151–152*
  – pinch–off areas *154, 155, 156*
– ejection of *159*
– finishes *157*
– materials *151*
  – berylliumcopper *151*
  – cast aluminum *151*
  – steel *151*
– parison, cutting and welding of *154–156*
  – pinch-off section *154–156*
– venting, importance of *157, 158, 159*
– weld lines, uniformity of *156*
– *See also* Injection blow molding; Injection blow molding, process of
Molding *1*
Mold rotation system, schematic of *9*

Morphology *186*
Multiple/combination cavities *66*

Neck ring
– design of *159–161*
– insert *165–166, 167*
NEPCO *58*
Networks, computer *175–176*
Newtonian fluid, blow molding simulation of *76–78*
Northern Engineering and Plastic Corporation. *See* NEPCO

OBG design *144*
Overflow weight standards *52–53*
Owens *2*

Pad printing. *See* Decoration of products
Parison
– cutting and welding of *154–156*
– formation of *1*
– programming of *117*
– *See also* Computer software; Bottles, design for
Parts, design of
– basic considerations
– hollow structures, understanding of *27–28*
– part, draft of *29–30*
– size *27, 28*
– personnel *28*
Parts, routing of *179*
Parts, scheduling of *179, 180*
PD2MS. *See* Product Design and Development Management System
Peel-off label *91, 92*
Performance criteria *73*
Personal computers *174–175*
PERT *17*
Pin adapter *114*
PITA. *See* Polymer Inflation and Thinning Analysis
Placo X-Y machine *135–138*
Polyflow (software program), *76, 78–79*
Preform. *See* Parison
Plastic bottles, design of *4*
Plastic molding processes, comparison of *10*
Plasticizing *1*
Plastics. *See* Polymers

Plastics, engineered 5–6
Plax Corporation 2, 3
Polyethylene, high–density 4, 5
– products, characteristics of 5
Polyethylene screw 105
Polymer Inflation and Thinning Analysis (PITA) (software program), 80–83
– Polymers 181, 182–183
 – chemistry of 181–183
  – matter, structure of 181–183
  – properties of 184–187
  – types of 183–184
– coloring of 196–197
– parameters of 190
– physical properties of 187–190
  – creep 190
  – hardness 189
  – melt flow rate 188
  – moisture 188–189
  – specific gravity 187–188
  – tensile strength 189–190
– regrinding 197198
– See also Recycled materials
Polymers, blow molding of
– acrylonitrile-butadiene-styrene 191, 193
– polyethylene, high-density 191, 192
– polyethylene terephthalate 196
– polycarbonate 193–194
– polyphenylene oxide 195–196
– polypropylene 194–195
– See also Blow molding; Blow molding, process of
"Precedence" diagramming 17
Pressure label 91, 92
Pressure ring 114
Primitive shapes 80
Product Design and Development Management System 16
– concurrent engineering 19
– phases of 16–17
  – commercialization 22
  – design 17
  – idea 16
  – implementation 22–23
  – production 25
  – test 24
– project management systems 17
– resources, commitment of 17–23

Program Evaluation and Review Technique. See PERT

Quality assurance 179

Radii, guidelines for 33–36
– corners and edges 35–36
Reaction injection molding (RIM), 60
Reciprocating screw machine 104
Recycled materials 198–199
Reed-Prentice injection molding machine 4
Reinforcing agents 186–187
Resins
– growth of market in 6–8, 9
– hydroscaopic 188–189
Resin/fiberglass layup 60
Ribs and gussets 36–37
– horizontal 47
– structural 60
– vertical 47
RIM. See Reaction injection molding
Rising mold 109
Robotic equipment 1
Rotary wheel 109–110
Rotating molds 163
Rotational molding 8, 9, 10
Rotec valve 4, 5

Sanding 90
Screen printing. See Decoration of products
Screws
– barrier 105
– extruder 104–105
– polyethylene 105
– reciprocating 110
Shuttle system 108–109
Simulation process 74
Sliding–bottom molds 163
Snap fits 65, 66
Society of the Plastic Industry guidelines 48, 49, 50, 58
Solvents 88
Specific gravity 187–188
Spray painting. See Decoration of products
Stacking angle 69
Steel molds 151
Stiffness, molded–in 60
Stretch blow molding 124, 125

Surface treatment. *See* Decoration of products

Tensile strength *189–190*
Terpolymers *183*. *See also* Polymers
Thermoforming *9, 10, 28*
Thermoplastic polymers *183*
– crystalline *183–184*
– amorphous *183–184*.
– *See also* Polymers
Termoset polymers *183*. *See also* Polymers
Three-dimensional blow molding *135, 136*
– curved *141*
– foam technology *144–145*
– geometry of *77–78*
– hard-soft-hard and soft-hard-soft technology *141–142*
  – axial coextrusion *141, 142*
  – material combinations, preferred *141–142*
– long-glass-fiber reinforced *143*
– Placo X-Y machine *136, 137, 138*
  – features of *137, 138*
  – formed parts *136–137*
  – X-Y Process *136–137*

– suction *139–140*
– *See also* Blow molding; Blow molding, process of
Truss groove *47*
– Tube. *See* Parison

Uniloy. *See* Johnson Controls

Vapor degreasing *90*
Vents *167–168*
– importance of *157, 158, 159*
Volume capacity and measurement *48–49, 51–58*

Waldron–Hartig. *See* Davis Standard
Wall thickness *9, 75, 80–83*
– distribution curves *80–83*
– and extrusion blow molding *116–118*
Warpage, in containers *66–67*
Water-based chemicals, washing with *88*
Weld lines, uniformity of *156*

Zarn, Inc. *4*